Essential Quotes for Scientists and Engineers

Konstantin K. Likharev

Editor

Essential Quotes
for Scientists and Engineers

 Springer

Editor
Konstantin K. Likharev
South Setauket, NY, USA

ISBN 978-3-030-63331-8   ISBN 978-3-030-63332-5   (eBook)
https://doi.org/10.1007/978-3-030-63332-5

This Springer imprint is published by the registered company Springer Nature Switzerland AG
The registered company address is: Gewerbestrasse 11, 6330 Cham, Switzerland

*It is a good thing for an uneducated man
to read books of quotations.*

—*Winston Churchill, 1930*

Preface

For following Sir Winston's advice taken for the epigraph, there are many quote collections available, both in print and online—see, e.g., the list of references at the end of this publication. However, I could not find a collection that would be, to my taste, reasonably comprehensive, at the same time being sufficiently refined for a typical busy member of the science/engineering community, to which I belong. This is why I dare to offer this community (including students of STEM disciplines) a modest set of less than 2500 my personal favorites, accumulated over nearly a decade of bedtime reading.

These quotes (on quite a few topics, not just science and engineering per se) have been selected for either their importance for the community, or depth of thought, or wit—or all of the above. Any selection implies a certain bias, but this collection still includes different perspectives—in my humble view, all non-trivial —on quite a few debatable issues.

To make consecutive reading of the quotes more engaging, the entries are not only grouped into topical sections, but also thematically ordered inside each section, forming quasi-continuous narrative threads. The unavoidable twists and turns and (less frequent) breaks of the threads are marked with three-asterisk lines.

The text and the authorship of each quote have been carefully verified.[1] If the time of its first publication (sometimes posthumous) could be established, it is given after the author's name; if not, some idea of its timing may be obtained from the years of author's life, listed in the *Author Index*, and also from my footnote comments. Since most of these comments address the authorship and timing issues, they may be safely ignored by the reader interested only in the quote contents.

South Setauket, NY, USA   Konstantin K. Likharev

---

[1]An entry was included into the final set only if it had been found either in the original publication, or in at least two reliable quote collections. (Of the online sources I had explored, I held only *Libquotes*, *Quote Investigator*, and *Wikiquotes* in this category).

Contents

Primer: On Quotations

A short saying often contains much wisdom.

—Sophocles, fifth century BC

I quote others only in order to express myself better.

—Michel de Montaigne, 1595

The devil can cite Scripture for his purpose.

—William Shakespeare, 1596

The next best thing to being witty oneself, is to be able to quote another's wit.[1]

—Christian Nestell Bovee, 1862

When reading a scholarly critic, one profits more from his quotations than from his comments.

—W. H. Auden, 1962

That's the point of quotations, you know: one can use another's words to be insulting.

—Amanda Cross, 1971

Quotation, *n*. The act of repeating erroneously the words of another.

—Ambrose Bierce, 1911

Misquotation is, in fact, the pride and privilege of the learned.

—Hesketh Pearson, 1934

---

[1] Of several later paraphrases of this quip, perhaps the most popular is one by W. Somerset Maugham: "The ability to quote is a serviceable substitute for wit".

K. K. Likharev (ed.), *Essential Quotes for Scientists and Engineers*,
https://doi.org/10.1007/978-3-030-63332-5_1

As many people, so many pronouncements.[2]

—Terence, ∼ 160 BC

Quotation confesses inferiority.

—Ralph Waldo Emerson, 1876

I always have a quotation for everything – it saves original thinking.

—Dorothy L. Sayers, 1932

A witty saying proves nothing.

—Voltaire, 1767

---

[2]The original Latin sounds even better: "*Quot homines, tot sententiae*".

On Discoveries, Problems, and Solutions

All truths are easy to understand once they are discovered; the point is to discover them.

—Galileo Galilei, 1632

The more original a discovery, the more obvious it seems afterwards.

—Arthur Koestler, 1970

The problem is not so much to see what nobody has yet seen, as to think what nobody has yet thought.

—Arthur Schopenhauer, 1851[1]

It isn't that they can't see the solution. It is that they can't see the problem.

—G. K. Chesterton, 1935

To see what is in front of one's nose needs a constant struggle.

—George Orwell, 1946

There is no regular procedure, no logical system of discovery, no simple, continuous development.

—Gerald Holton, 1973

There are many rules for choosing good solutions, but there are no rules for selecting good rules.

—Anonymous

***

It requires a very unusual mind to undertake the analysis of the obvious.

—Alfred North Whitehead, 1925

Talent hits a target no one else can hit; genius hits a target no one else can see.

—Arthur Schopenhauer, 1844

---

[1]Frequently misattributed to Albert Szent-Györgyi, who just quoted it in 1957.

© The Author(s), under exclusive license to Springer Nature Switzerland AG 2021
K. K. Likharev (ed.), *Essential Quotes for Scientists and Engineers*,
https://doi.org/10.1007/978-3-030-63332-5_2

In every work of genius we recognize our own rejected thoughts; they come back to us with a certain alienated majesty.

—*Ralph Waldo Emerson, 1841*

From an individual exceptional man alone is any great addition to be expected to a sum of original conceptions. Who could imagine Faraday as cooperating!?

—*Theodore William Richards, 1902*

One of the greatest pains to human nature is the pain of a new idea.

—*Walter Bagehot, 1884*

The human likes a strange idea as little as the body likes a strange protein, and it resists it with a similar energy.

—*Wilfred Trotter, 1941*[2]

Don't worry about people stealing an idea. If it's original, you will have to ram it down their throats.

—*Howard H. Aiken, 1987*

We don't believe anything we don't want to believe.

—*Theodore Sturgeon, 1953*

It takes years to drive an idea through a quarter-inch of human skull.

—*Charles F. Kettering, 1961*

A new scientific truth does not triumph by convincing its opponents and making them see the light, but rather because its opponents eventually die, and a new generation grows up that is familiar with it.

—*Max Planck, 1950*[3]

Whenever a new and startling fact in brought to the light in science, people first say, 'it is not true', then 'it is contrary to religion', and lastly, 'everybody knew that before'.

—*Louis Agassiz*[4]

***

An ingenious conjecture greatly shortens the road.

—*Gottfried Leibniz, 1685*

Intuition isn't the enemy, but the ally, of reason.

—*John Kord Lagemann, 1955*

---

[2]A later paraphrase of this quote by P. B. Medawar is even more popular.

[3]A popular shorter (anonymous?) paraphrase of this aphorism is: "Science advances one funeral at a time".

[4]As quoted in 1863 by Charles Lyell. This famous maxim, possibly based on an earlier (circa 1819) line by Arthur Schopenhauer, was paraphrased later by William James (in 1907), W. I. B. Beveridge (in 1955), J. B. S. Haldane (in 1963), Arthur C. Clarke (in 1968), Dorothy Maud Wrinch, and probably many others.

Every great advance in science had issued from a new audacity of imagination.

—*John Dewey*, 1929

The scientist, if he is to be more than a plodding gatherer of bits of information, needs to exercise an active imagination.

—*Linus Pauling*, 1943

Imagination comes first in both artistic and scientific creations, but in science there is only one answer and that has to be correct.

—*James Watson*, 1987

Every now and then, a man's mind is stretched by a new idea or sensation, and never shrinks to its former dimensions.

—*Oliver Wendell Holmes Sr.*, 1858

Lack of money is no obstacle. Lack of an idea is an obstacle.

—*Ken Hakuta*, 1988

Be less curious about people and more curious about ideas.

—*Marie Curie*[5]

If you want to have good ideas you mush have many ideas. Most of them will be wrong, and what you have to learn is which ones to throw away.

—*Linus Pauling*, 1995

There are well-dressed foolish ideas - just as there are well-dressed fools.

—*Nicolas Chamfort*[6]

For an idea ever to be fashionable is ominous, since it must afterwards be always old-fashioned.

—*George Santayana*, 1913

Whenever sleep seized me I would see those very problems in my dreams; and many questions became clear to me in my sleep.

—*Avicenna*[7]

No one has ever had an idea in a dress suit.

—*Frederick Banting*, 1922

Beware of all enterprises that require new clothes.

—*Henry David Thoreau*, 1854

Be regular and orderly in your life [...], so that you may be violent and original in your work.

—*Gustave Flaubert*, 1876

---

[5]As quoted, without a date, by E. Curie Labouisse in 1937.

[6]As quoted, without a date, by William G. Hutchinson in 1902.

[7]As quoted, without a date, by W. E. Goldman in 1974.

Do not fear to be eccentric in opinion, for every opinion now accepted was once eccentric.

—*Bertrand Russell*, 1951

I'll be more enthusiastic about encouraging thinking outside the box when there's evidence of any thinking going on inside it.

—*Terry Pratchett*, 2016

***

A beautiful idea has a much greater chance of being correct than a ugly one.

—*Roger Penrose*, 1989

If nature were not beautiful it would not be worth knowing, and [then] life would not be worth living.

—*Henri Poincaré*, 1908

Nature is never so admired as when she is understood.

—*Bernard Le Bovier de Fontenelle*, 1733

***

There are men who never err, because they never propose anything rational.

—*Johann Wolfgang von Goethe*, 1832

The man who makes no mistakes does not usually make anything.[8]

—*E. J. Phelps*, 1889

You must never feel badly about making mistakes [...] as long as you take the trouble to learn from them.[9]

—*Norton Juster*, 1961

An error doesn't become a mistake until you refuse to correct it.

—*O. A. Battista*, 1946

Who excuses himself, accuses himself.

—*European proverb*[10]

Man is condemned to exhaust all possible errors when he examines any set of facts before he recognizes the truth.

—*Jean-Baptiste Lamarck*, 1809

---

[8]In his full text, Phelps explicitly refers to an earlier (circa 1868) maxim by Archbishop William Connor Magee. Later paraphrases by Joseph Conrad, Theodore Roosevelt, and Marva Collins are also frequently quoted.

[9]See also the second quote in the *On the Wise...* section.

[10]Here and below, the label *European* is used for any proverb that is similar (within a reasonable flexibility of translation) in two or more European languages.

Research is the process of going up alleys to see if they are blind.

—*Marston Bates*, 1950

For men with insatiable curiosity, their prototype is [...] the dog sniffing tremendously at an infinite series of rat-holes.

—*H. L. Mencken*, 1923

It is a good morning exercise for a research scientist to discard a pet hypothesis every day before breakfast. It keeps him young.

—*Konrad Lorenz*, 1966

There is no failure except in no longer trying.

—*Elbert Hubbard*, 1895

You may be disappointed if you fail, but you are doomed if you don't try.

—*Beverly Sills*, 1997

Our greatest glory is not in never falling, but in rising every time we fall.

—*Confucius*, fifth century BC

Fall down seven times, get up eight.

—*Japanese proverb*[11]

If you want to succeed, double your failure rate.

—*Thomas J. Watson*[12]

Only those who dare to fail greatly can ever achieve greatly.

—*Robert F. Kennedy*, 1966

Success teaches us nothing; only failure teaches.

—*Hyman G. Rickover*, 1954

Good judgment comes from experience. Experience comes from bad judgment.

—*Anonymous*[13]

***

He who risks nothing, can gain nothing.

—*Italian proverb*

He who does not risk, does not drink champagne.

—*Russian proverb*

To win without risk is to triumph without glory.

—*Pierre Corneille*, 1636

---

[11]Here and below, proverbs are labeled by the languages in which they are common, rather than their countries of origin.

[12]As quoted, without a date, by Roger von Oech in 1982.

[13]I have seen this joke attributed to many authors, including Will Rogers, Rita Mae Brown, Randy Pausch, and even Mullah Nasreddin Hooja, but could not establish the actual original authorship.

He who knows how to endure everything can risk everything.

—*Luc de Clapiers*, 1746

All things are difficult before they are easy.

—*Thomas Fuller*, 1732

They can because they think they can.

—*Virgil*[14]

Where there is a will, there is a way.

—*European proverb*

Will is power.

—*French proverb*

Nothing great in the world has been accomplished without passion.

—*Georg Wilhelm Freidrich Hegel*, 1821

Expect problems and eat them for breakfast.

—*Alfred Armand Montapert*, 1970

We are continually faced with a series of great opportunities brilliantly disguised as insoluble problems.

—*John W. Gardner*, 1966

Men are generally idle and ready to satisfy themselves and intimidate the industry of others by calling that impossible which is only difficult.

—*Samuel Johnson*, 1787

So many things are possible just as long as you don't know they're impossible.

—*Norton Juster*, 1961

Things that are merely difficult never give lasting pleasure.

—*Voltaire*, 1759

Madam, if a thing is possible, consider it done; the impossible? that will be done.

—*Charles Alexandre de Calonne*, 1783

The greatest pleasure in life is doing what other people say you cannot do.

—*Walter Bagehot*, 1853

The only way of finding the limits of impossible is by going beyond them to the impossible.

—*Arthur C. Clarke*, 1972

Things are only impossible until they're not.

—*Hannah Louise Shearer*, 1988

---

[14]From his *Aeneid*, written in 29-19 BC.

It's kind of fun to do the impossible.

—Walt Disney[15]

The best way out is always through.

—Robert Frost, 1914

Where you cannot climb over, you must creep under.

—Danish proverb

Seek, and ye shall find.

—The Bible (Matthew 7:7)

To strive, to seek, to find, and not to yield.

—Alfred Tennyson, 1842

Where one door is shut, another opens.

—Spanish proverb

***

In the end, there is nothing too remote to be reached, or too well hidden to be discovered.

—René Descartes, 1637

We never stop investigating. [...] This has become the greatest survival trick of our species.

—Desmond John Morris, 1967

***

Patience is the greatest of all virtues.

—European proverb[16]

If I have ever done the public any service [...], 'tis due to nothing but industry and a patient thought.

—Isaac Newton, 1692

He that can have Patience, can have what he will.

—Benjamin Franklin, 1736

Genius is nothing else than a great aptitude for patience.

—George-Louis Leclerc, 1785

There is only one major sin: impatience.

—Franz Kafka, 1917

---

[15]As quoted, without a date, by Derek Walter in 1982.

[16]It is sometimes attributed to Cato, but I could not find a reliable contribution of his authorship.

Success in research needs four *G*s: *Glück, Geduld, Geschick, und Geld* (luck, patience, skill, and money).

—Paul Ehrenfest[17]

Let me tell you the secret that has led me to my goal. My strength lies solely in my tenacity.

—Louis Pasteur[18]

Toil conquered the world, unrelenting toil.

—Virgil[19]

No fine work can be done without concentration and self-sacrifice and toil and doubt.

—Max Beerbohm, 1914

The only factor becoming scarce in a world of abundance is human attention.

—Kevin Kelly, 1998

***

Dreams do come true, if we only wish hard enough. You can have anything in life if you will sacrifice everything else for it.

—J. M. Barrie, 1904

To gain that which is worth having, it may be necessary to lose everything else.

—Burnadette Devlin, 1969

We succeed only as we identify in life, or in war, or in anything else, a single overriding objective, and make all other considerations bend to that one objective.

—Dwight Eisenhower, 1957

***

Make no little plans; they have no magic to stir men's blood. [...] Make big plans, aim high in hope and work.

—Daniel Burnham, 1910[20]

Too low they build, who build beneath the stars.

—Edward Young[21]

For a big ship, a long sailing.[22]

—Russian proverb

---

[17]As quoted, without a date, by M. Perutz in 1988.

[18]As quoted, without a date, by Wayne W. Dyer in 2001.

[19]From his *Aeneid*, written in 29-19 BC.

[20]Reportedly, first put on paper by a Burnham's associate, some Willis Polk.

[21]From his *Night Thoughts*, written in 1742-1745.

[22]The recent (anonymous?) mocking, "For a big ship, a big torpedo", also seems valid not only literally—see the *On Society...* section below.

Most people would succeed in small things if they were not troubled with great ambitions.

—Henry Wadsworth Longfellow, 1857

The human tendency to regard little things as important has produced very many great things.

—Georg Christoph Lichtenberg[23]

If politics is the art of the possible, research is surely the art of the soluble.

—P. B. Medawar, 1964

Pick battles big enough to matter, small enough to win.

—Jonathan Kozol[24]

I have little patience with scientists who take a board of wood, look for its thinnest part and drill a great number of holes where drilling is easy.

—Albert Einstein, 1949

'Tis Ambition enough to be employed as an Under-Laborer in clearing Ground a little, and removing some of the Rubbish that lies in the way to Knowledge.

—John Locke, 1690

The most likely way to reach a goal is to be aiming not at that goal itself but at some more ambitious goal beyond it.

—Arnold J. Toynbee[25]

***

Simplicity is the peak of civilization.

—Jessie Sampter, 1927

Genius is the ability to reduce the complicated to the simple.

—C. W. Ceram, 1951

It's the Simple things that are really effective.

—Theodore Sturgeon, 1941

There is always a well-known solution to every human problem - neat, plausible, and wrong.

—H. L. Mencken, 1917

It was beautiful and simple, as all truly great swindles are.

—O. Henry, 1908

---

[23]From his *Notebook G* (1779-1783).

[24]As quoted, without a date, in several reputable collections, starting not later than 1981.

[25]As quoted, without a date, by Edward Myers in 1955. Note that this maxim is sometimes misattributed to the author's uncle, Arnold Toynbee (1852-1883).

The shortest distance between two points is under construction.

—Anonymous[26]

Seek simplicity, and distrust it.

—Alfred North Whitehead, 1919

I'm suspicious of anyone who has a strong opinion on a complicated issue.

—Scott Adams, 2007

Nothing is as simple as we hope it will be.

—Jim Horning[27]

The Lord God is subtle, though not malicious.

—Albert Einstein, 1946

I have yet to see any problem, however complicated, which, when you looked at it in the right way did not become still more complicated.[28]

—Poul Anderson, 1969

It's so much easier to suggest solutions when you don't know too much about the problem.

—Malcolm Forbes, 1978

– Not every problem has a good solution.
– Every solution has side effects.

—Dan Geer, 2007

If a problem has no solution, it may not be a problem, but a fact - not to be solved, but to be coped with over time.

—Shimon Peres, 2001

If a problem does not have a technical solution, it must not be a real problem. It is but an illusion […], a figment born of some regressive cultural tendency.

—Theodore Roszak, 1969

There is nothing more permanent than a temporary solution.

—Russian proverb[29]

***

Believe those who are seeking the truth. Doubt those who find it.

—André Gide, 1952

---

[26]This popular joke is frequently attributed to either Leo Aikman or some Noelie Altito, but I was unable to confirm either of these authorships, and even to identify the latter person.

[27]The (undated) title of one of his blogs.

[28]The so-called *Anderson's Law*.

[29]Adjusted (in 1980) by Milton Friedman to describe "temporary" government programs.

No human being is constituted to know the truth, the whole truth, and nothing but the truth; and even the best of men must be content with fragments, with partial glimpses, never the full fruition.

—William Osler, 1905

There are no whole truths; all truths are half-truths. It is trying to treat them as whole truths that plays the devil.[30]

—Alfred North Whitehead, 1954

Chase after the truth like all hell and you'll free yourself, even though you never touch its coat-tails.

—Clarence Darrow, 1938

***

A problem well stated is a problem half solved.

—Charles F. Kettering[31]

Well begun is half done.

—European proverb

***

Lead, follow, or get out of the way.

—Anonymous[32]

You're either part of the solution or you're part of the problem.

—Anonymous[33]

If you're not part of the solution, you're part of the precipitate.

—Henry J. Tillman[34]

---

[30]This is apparently a play around a much older line by François Rabelais: "Speak the truth and shame the devil".

[31]As quoted in several reliable sources, starting not later than 1954.

[32]This popular slogan (with slight variations) has been attributed to perhaps the record number of candidate authors, including Thomas Paine, Thomas Edison, George S. Patton, Ted Turner, Lee Iacocca, Laurence J. Peter, and Arthur Uther Pendragon. I was unable to establish the original author.

[33]This line and its close siblings, quoted in several newspapers in the 1950s and the 1960s as unauthored jokes, are actually succinct paraphrases of an earlier (circa 1937) maxim by John R. Alltucker. The line became famous after it had been used by Eldridge Cleaver in his 1968 speech.

[34]As quoted, without a date, by Robert A. Day and Nancy Sakaduski in 2011. Unfortunately I was unable to identify this author.

On Work and Leisure

Give me a place to stand, and I shall move the earth.

—*Archimedes*[1]

Give me a place to sit, and I shall watch.

—*Anonymous*[2]

I like work: it fascinates me. I can sit and look at it for hours.

—*Jerome K. Jerome*, 1889

Somebody has to do something, and it's just incredibly pathetic that it has to be us.

—*Jerry Garcia*, 1988

We can't all be heroes because somebody has to sit on the curb and clap as they go by.

—*Will Rogers*[3]

The reason why worry kills more people than work is that more people worry than work.

—*Robert Frost*, 1914

Science may have found a cure for most evils; but it has found no remedy for the worst of them all - the apathy of human beings.

—*Helen Keller*, 1927

The gods help them that help themselves.

—*Aesop*[4]

---

[1] As quoted by Pappus of Alexandria (in $\sim$340 AD).

[2] Sometimes "attributed" to an "Archimedes' friend".

[3] As quoted, without a date, by Jacob Morton Braude in 1969.

[4] From his fable *Hercules and the Wagoner*, written in the mid-fifth century BC.

K. K. Likharev (ed.), *Essential Quotes for Scientists and Engineers*,
https://doi.org/10.1007/978-3-030-63332-5_3

Nothing venture, nothing have.

—French proverb

Try to turn every disaster into an opportunity.

—John D. Rockefeller[5]

When opportunity knocks, open the door.

—European proverb

Opportunity is missed by most people because it is dressed in overalls and looks like work.

—Thomas Edison[6]

What the country needs is dirtier fingernails and cleaner minds.

—Will Rogers[7]

***

One cannot, in any shape or form, depend on human relations for lasting reward. It is only work that truly satisfies.

—Bette Davis, 1962

All work is noble; work is alone noble.

—Thomas Carlyle, 1843

The only place where success comes before work is in the dictionary.

—Anonymous[8]

It takes all the running you can do, to keep in the same place. If you want to get somewhere else, you must run at least twice as fast as that!

—Lewis Carroll, 1865

All that is human must retrograde if it do not advance.

—Edward Gibbon, 1776

He who does not move forward, goes backward.

—Johann Wolfgang von Goethe, 1797

The world belongs to the energetic. Every great and commanding moment in the annals of the world is the triumph of some enthusiasm.

—Ralph Waldo Emerson, 1876

---

[5]As quoted, without a date, by Peter Collier and David Horowitz in 1937.

[6]As quoted, without a date, by John L. Mason in 1990.

[7]As quoted, without a date, by A. S. Migs Damiani in 1998.

[8]This maxim, whose unauthored versions may be found in press at least as early as in 1925, was later paraphrased by many, including Vince Lombardi.

Nature is entirely neutral; she submits to him who most energetically and resolutely assaults her. She grants her rewards for the fittest.

—William Graham Sumner, 1880

Act [...] when and where you want to, but do it with all your might.

—J. D. Salinger, 1957

***

Whatever women do they must do twice as well as men to be thought half as good. Luckily, this is not difficult.

—Charlotte Whitton, 1963

Sure [Fred Astaire] was great, but don't forget that Ginger Rogers did everything he did... backwards, and in high heels!

—Bob Thaves, 1982

Women who seek to be equal with men lack ambition.

—Timothy Leary, 1987

[A woman] needs to know how to look like a girl, act like a lady, think like a man, and work like a dog.

—Caroline K. Simon, 1959

***

A certain amount of opposition is a great help to a man. Kites rise against, not with, the wind.

—John Neal, 1848

To reach the source, one has to swim against the current.

—Stanisław Jerzy Lec, 1962

Only dead fish swims with the stream.

—Malcolm Muggeridge, 1966

What does not kill me, makes me stronger.

—Friedrich Nietzsche, 1888

A pike is in the sea to keep a crucian alert.

—Russian proverb

Fortune helps the bold.

—Virgil, ~ 19 BC

Diligence is the mother of good fortune.

—Miguel de Cervantes, 1615

No victor believes in chance.

—Friedrich Nietzsche, 1882

Where observation is concerned, chance favors only the prepared mind.

—Louis Pasteur, 1854

The winds and waves are always on the side of the ablest of navigators.

—Edward Gibbon, 1776

Readiness is the mother of luck.

—Baltasar Gracián, 1647

I'm a great believer in luck, and I find the harder I work, the more I have of it.

—Stephen Leacock, 1910[9]

The only thing that overcomes hard luck is hard work.[10]

—Harry Golden, 1955

***

We should be taught not to wait for inspiration to start a thing.

—Frank Tibolt[11]

Inspiration does exist, but it must find you working.

—Pablo Picasso[12]

Aspiration creates inspiration.

—Edward Abbey, 1990

Genius is one percent inspiration, ninety-nine percent perspiration.

—Thomas Edison, 1902[13]

Sweat makes good mortar.

—German proverb

***

Great things are not done by impulse, but by a series of small things brought together.

—Vincent van Gogh, 1882

The steady drip of water causes stone to hollow and yield.

—Lucretius, first century BC

---

[9]Frequently misattributed to Thomas Jefferson.

[10]This advice looks problematic: for most people, hard work is also hard luck.

[11]From his book *A Touch of Greatness*, whose 2nd edition was published in 1981. (The first one was published either in the 1940s or in the 1950s.)

[12]As quoted, without a date, by Tomas R. Villasante in 1994.

[13]Starting from 1898, Edison offered several versions of this famous line, with the "inspiration" fraction alternating between one and two percent.

You can go a long way after you have got tired.

—French proverb

By perseverance the snail reached the ark.

—Charles Haddon Spurgeon, 1869

Perseverance is a great element of success. If you only knock long enough and loud enough at the gate, you are sure to wake up somebody.

—Henry Wadsworth Longfellow, 1887

It is always darkest just before the dawn.

—Thomas Fuller, 1650

Everything has an end – except a sausage, which has two.

—Danish proverb

It is pleasant to remember what was hard to endure.

—Italian proverb

The greater the effort, the greater the glory.

—Pierre Corneille, 1642

Perhaps the most valuable result of all education is the ability to make yourself do the thing you have to do, when it ought to be done, whether you like it or not.

—Thomas Henry Huxley, 1877

Wisdom is knowing what to do next, [...] virtue is doing it.[14]

—David Starr Jordan, 1898

Who will not when he can, cannot when he will.

—European proverb

Do what you can, with what you've got, where you are.

—Bill Widener[15]

***

Whatever is worth doing at all, is worth doing well.

—Lord Chesterfield, 1746[16]

---

[14]There are several later paraphrases of Jordan's first sentence, along the following line: "The only practical question in life is: what to do next?".

[15]Often misattributed to Theodore Roosevelt, who just quoted this maxim (with a reverent reference to Widener) in 1913.

[16]Possibly based on the Latin maxim *"Age Quod Agis"*, literally meaning "Do what you are doing".

Quality is never an accident; it is always the result of intelligent effort.

—*John Ruskin*, 1853

It is just the little touches after the average man would quit that make the master's fame.

—*Orison Swett Marden*[17]

People forget how fast you did a job - but they remember how well you did it.

—*Howard W. Newton*[18]

Quick and well seldom go together.

—*European proverb*

Perfection is the dream of imperfection that refuses to wake up.

—*Fausto Cercignani*, 2004

Good is the enemy of perfect.[19]

—*European proverb*

If you want a thing done well, do it yourself.

—*European proverb*[20]

***

Take everything you like seriously, except yourselves.

—*Rudyard Kipling*, 1928

There are people who think that everything is sane and sensible that is done with solemn face.

—*Georg Christoph Lichtenberg*[21]

You must not think me necessarily foolish because I am facetious, nor will I consider you necessarily wise because you are grave.

—*Sydney Smith*, 1840

You've achieved success in your field when you don't know whether what you're doing is work or play.

—*Warren Beatty*, 2004

Play by the rules, but be ferocious.

—*Bill Bowerman*[22]

---

[17]Included, without a date, into all recent collections of his works.

[18]As quoted, without a date, in several reliable sources, starting not later than 1945.

[19]Not very surprisingly, the reversed version of this proverb is even more popular.

[20]This may be just a popular paraphrase of a longer line by Heinrich Bullinger.

[21]From his *Notebook E* (1775-1776).

[22]Frequently misattributed to Phil Knight, who just quoted (with a reverent reference) Bowerman, his own athletic coach and business partner.

To play it safe is not to play.

—Robert Altman[23]

A ship in harbor is safe – but that is not what ships are built for.

—John Augustus Shedd, 1928[24]

To love what you do and feel that it matters – how could anything be more fun?

—Katharine Graham, 1994

Real joy comes not from ease or riches or from the praise of men, but from doing something worthwhile.

—Wilfred Grenfell[25]

The only thing that may bring joy is work.

—Louis Pasteur, 1940

Far and away the best prize that life offers is the chance to work hard at work worth doing.

—Theodore Roosevelt, 1903

Choose a job you love, and you will never have to work a day in your life.

—Anonymous[26]

I never did a day's work in my life. It was all fun.

—Thomas Edison[27]

The test of a vocation is the love of the drudgery it involves.

—Logan Pearsall Smith, 1931

***

Too much rest is rust.

—Walter Scott, 1825[28]

Rest's for a clam in a shell.

—Dorothy Parker, 1944

The prospect of a long day at the beach makes me panic.

—Phillip Lopate, 1994

A perpetual holiday is a good working definition of hell.

—George Bernard Shaw, 1910

---

[23] As quoted, without a date, in several generally reliable collections, starting not later than 1978.

[24] Sometimes misattributed to William Greenough Thayler Shedd (1820-1894).

[25] As quoted, without a date, in several reputable sources, starting not later than 1967.

[26] Frequently misattributed to Confucius.

[27] As quoted, without a date, in a 2003 collection published by *Edison and Ford Winter Estates*.

[28] Actually, Scott represented that maxim as a line from an "old song".

No man needs a vacation so much as the man who has just had one.

—Elbert Hubbard, 1904

Idleness attracts vices.

—Saint Bernard[29]

Idle man does not know what it is to enjoy rest, for he has not earned it.

—John Lubbock, 1913

An idle brain is the devil's workshop.

—English proverb

Blessed is the person who is too busy to worry in the daytime and too sleepy to worry at night.

—Anonymous[30]

Man is so made that he can only find relaxation from one kind of labor by taking up another.

—Anatole France, 1881

A change is as good as a rest.

—English proverb

***

Nobody who's really using his ego, his real ego, has any time for any goddam hobbies.

—J. D. Salinger, 1957

It is impossible to imagine Goethe or Beethoven being good at billiards or golf.

—Anonymous[31]

To be able to fill leisure intelligently is the last product of civilization, and at present very few people have reached this level.

—Bertrand Russell, 1930

There is no public entertainment which does not inflict spiritual damage.

—Tertullian, ~ 197 AD

Pleasures steal away the mind.

—Dutch proverb

---

[29]As quoted, without a date, by Geoffrey Chaucer in his *Canterbury Tales* (written between 1387 and 1400).

[30]I have seen this aphorism attributed to Leo Aikman, but could not find a reliable confirmation of his authorship.

[31]Broadly (and very plausibly) attributed to H. L. Mencken, but I was unable to find a reliable confirmation of his authorship.

Amusement is the happiness of those who cannot think.

—Alexander Pope, 1727

Every useless amusement is an evil for a being whose life is so short and whose time is so pressing.

—Jean-Jacques Rousseau, 1762

Everything considered work is less boring than amusing oneself.

—Charles Baudelaire, 1864

When men are rightly occupied, their amusement grows out of their work.

—John Ruskin, 1865

Free time is an illusion. It's what you get when you die and the gods reward you for a life spent working from dawn until midnight.

—Tamora Pierce, 1983

On Science and Technology

There is more love for humanity in electricity and steam than in chastity and abstinence from meat.[1]

—*Anton Chekhov*, 1894

Those [changes in history] that have proved permanent—the ones that affected every facet of life and made certain that mankind could never go back again—were always brought about by science and technology.

—*Isaac Asimov*, 1980

The scientist and engineer have built the modern world, and they hold the key to its control and coordination.

—*Benjamin C. Gruenberg*, 1935

The lack of ideas and inventions in one generation can easily mean the loss of freedom in the next.

—*Charles F. Kettering*, 1955

Science and technology multiply around us. To an increasing extent they dictate the languages in which we speak and think. Either we use those languages, or we remain mute.

—*J. G. Ballard*, 1984

We live in a society exquisitely dependent on science and technology, in which hardly anyone knows anything about science and technology.

—*Carl Sagan*, 1989

History without the history of science [...] resembles a statue of Polyphemus without his eye.

—*I. Bernard Cohen*, 1952

***

---

[1]This is a reference to Leo Tolstoy's moral teachings that were very popular in Russia of that time.

© The Author(s), under exclusive license to Springer Nature Switzerland AG 2021
K. K. Likharev (ed.), *Essential Quotes for Scientists and Engineers*,
https://doi.org/10.1007/978-3-030-63332-5_4

The progress of mankind is due exclusively to the progress of natural sciences, not to morals, religion, or philosophy.

—*Justus von Liebig*, 1866

Scientific thought is not an accompaniment or condition of human progress, but human progress itself.

—*William Kingdon Clifford*, 1872

The acquisition and systematization of positive knowledge are the only human activities which are truly cumulative and progressive.

—*George Sarton*, 1936

Science is the only human activity that is truly progressive.

—*Edwin Powell Hubble*, 1936

Science is the highest personification of the nation because that nation will remain the first which carries the furthest the works of thought and intelligence.

—*Louis Pasteur*, 1876

Science is much more than a body of knowledge. It is a way of thinking.

—*Carl Sagan*, 1989

Science is the acceptance of what works and the rejection of what does not. That needs more courage than we might think.

—*Jacob Bronowski*, 1951

Science is the great antidote to the poison of enthusiasm and superstition.

—*Adam Smith*, 1776

***

Science [consists of] the most complete possible presentation of facts with the least possible expenditure of thought.

—*Ernst Mach*, 1893

A science is any discipline in which the fool of this generation can go beyond the point reached by the genius of the last generation.

—*Max Gluckman*, 1965

If I have seen further it is by standing on the shoulders of Giants.

—*Isaac Newton*, 1675[2]

---

[2]This famous statement is actually a self-depreciating twist of an earlier (circa 1126) maxim by Saint Bernard: "We are like dwarfs perched on the shoulders of giants, and thus are able to see more and farther than they did", which is possibly based on an even earlier ($\sim$500 AD) pronouncement by Latin grammarian Priscian.

In the sciences, we are now uniquely privileged to sit side-by-side with the giants on whose shoulders we stand.

—Gerald Holton, 1961[3]

***

The whole of science is nothing more than the refinement of everyday thinking.

—Albert Einstein, 1936

The scientific method is a potentiation of common sense.

—P. B. Medawar, 1969

Common sense in an uncommon degree is what the world calls wisdom.

—Samuel Taylor Coleridge, 1827

Science is nothing but [...] common sense rounded up and minutely articulated.

—George Santayana, 1906

Science is nothing but trained and organized common sense, differing from the latter only as a veteran may differ from a raw recruit.

—Thomas Henry Huxley, 1894

Common sense, however it tries, cannot avoid being surprised from time to time. The aim of science is to save it from such surprises.

—Bertrand Russell, 1987

Science is a first-rate piece of furniture for a man's upper chamber, if he has common sense on the ground-floor.

—Oliver Wendell Holmes Sr., 1872

A weak mind with no common sense magnifies trifling things and cannot receive great ones.

—Anonymous[4]

Common sense in matters medical is rare, and is usually in inverse ratio to the degree of education.

—William Osler, 1894

Common sense is not so common.

—Anonymous[5]

---

[3] A popular (anonymous?) paraphrase says: "...to rub the shoulders with the giants on whose shoulders we stand".

[4] Frequently misattributed to Lord Chesterfield.

[5] This wisdom is frequently attributed to Voltaire, who indeed wrote a very close line, "Common sense is very rare", in his 1767 essay *Common Sense*, but only with the remark "People say sometimes...".

This is an age in which one cannot find common sense without a search warrant.

—George Will, 1996

***

Experience never errs; it is only your judgments that err by promising themselves effects such as are not caused by your experiments.

—Leonardo da Vinci[6]

We ought, in every instance, to submit our reasoning to the test of experiment, and never search for truth but by the natural road of experiment and observation.

—Antoine Lavoisier, 1790

An experiment is a device to make Nature speak intelligibly. After that one has only to listen.

—George Wald, 1968

The physicist can never subject an isolated hypothesis to experimental test, but only a whole group of hypotheses.

—Pierre Duhem, 1906

Experiment is the sole source of truth. It alone can teach us something new; it alone can give us certainty.

—Henri Poincaré, 1902

String theory is the first science in hundreds of years to be pursued in pre-Baconian fashion, without any adequate experimental guidance.

—Philip Warren Anderson, 2005

In God we trust. All others must show their data.

—Anonymous[7]

Facts speak for themselves.

—Terence, ∼160 BC

Plain matters of fact are terrible stubborn things.

—Eustace Budgell, 1732

Facts do not cease to exist because they are ignored.

—Aldous Huxley, 1927

In science 'fact' can only mean 'confirmed to such a degree that it would be perverse to withhold provisional assent'.

—Stephen Jay Gould, 1983

---

[6]Undated entry in his notebooks, as translated by J. P. Richter in 1883.

[7]Attributed to several authors, but the earliest of them, Edwin R. Fisher, disclaimed the credit, saying that at the time of his publication (1978) this was already a well-known cliché.

Where facts are few, experts are many.

—*Donald R. Gannon*[8]

Facts are the air of a scientist. Without them you never can fly. Without them your 'theories' are vain efforts.

—*Ivan Pavlov, 1936*

Any clod can have the facts, but having opinions is an art.

—*Charles McCabe*[9]

***

The trouble with facts is that there are so many of them.

—*Samuel McChord Crothers, 1903*

Science is built up of facts, as a house is built of stones, but a collection of facts is no more a science that a heap of stones is a house.

—*Henri Poincaré, 1902*

The important thing in science is not so much to obtain new facts as to discover new ways of thinking about them.

—*Lawrence Bragg, 1958*[10]

Theory is the essence of facts. Without theory, scientific knowledge would be only worthy of the madhouse.

—*Oliver Heaviside, 1893*

There is nothing more practical than a good theory.

—*Kurt Lewin, 1943*

In theory, there is no difference between theory and practice, while in practice, there is.

—*Benjamin Brewster, 1882*[11]

Today's scientists have substituted mathematics for experiments, and they wander off through equation after equation, and eventually build a structure which has no relation to reality.[12]

—*Nikola Tesla, 1934*

---

[8]Unfortunately, I could not identity this author, though the attribution is very common, starting not later than 1996.

[9]Apparently, first published in the *San Francisco Chronicle* some time between the mid-1950s and 1983.

[10]Sometimes misattributed to his farther, Sir William Henry Bragg (1862–1942).

[11]Frequently misattributed to some Chuck Reid, and less often to Albert Einstein and Richard Feynman.

[12]It is commonly acknowledged now that even such an inventor of genius as Tesla could benefit from using more math/theory in his work.

Who could believe an ant in theory?
a giraffe in blueprint?
Ten thousand doctors of what's possible
could reason half the jungle out of being.

—John Ciardi, 1969

The farther is an experiment from theory, the closer it is to the Nobel Prize.

—Irène Joliot-Curie[13]

I could trust a fact and always cross-question an assertion.

—Michael Faraday, 1858

A fact is a simple statement that everyone believes. It is innocent, unless found guilty. A hypothesis is a novel suggestion that no one wants to believe. It is guilty, until found effective.

—Edward Teller, 1991

From a closer and purer league between these two facilities, the experimental and the rational, [...] much may be hoped.

—Francis Bacon, 1620

The understanding cannot perceive; the senses cannot think. Only through their union can knowledge arise.

—Immanuel Kant, 1781

***

The art in scientific thinking—whether in physics, biology, or economics—is deciding which assumptions to make.

—N. Gregory Mankiw, 1998

Pick your assumptions to pieces till the stuff they are made of is exposed to plain view.

—Eric Temple Bell, 1935

Model-making [...] may be judged the most valuable part of scientific method because skill and insight in these matters are rare.

—Herbert George Andrewarta, 1961

A theory has only the alternative of being right or wrong. A model has a third possibility: it may be right but irrelevant.

—Manfred Eigen, 1973

One of the most insidious and nefarious properties of scientific models is their tendency to take over, and sometimes supplant, reality.

—Erwin Chargaff, 1978

***

---

[13]As quoted, without a date, by M. Markov in 1963.

The history of science teaches only too plainly the lesson that no single method is absolutely to be relied upon.

—Lord Rayleigh, 1882

Imposing an alleged uniform general method upon everybody breeds mediocrity in all but the very exceptional.

—John Dewey, 1916

We should not be ashamed to change our methods; rather we should be ashamed never to do so.

—Charles V. Chapin, 1924

All methods enabling substantial advance toward the truth are grade-A methods.

—I. I. Blekhman et al., 1983

It is in the works of application that one must study [methods]; one judges their utility there and appraises the manner of making use of them.

—Joseph-Louis Lagrange[14]

He who seeks for methods without having a definite problem in mind seeks for the most part in vain.

—David Hilbert, 1902

***

There is no more potent antidote to the corroding influence of mammon than the presence in the community of a body of men devoted to science, living for investigation and caring nothing for the lust of the eyes and the pride of life.

—William Osler, 1892

The power and salvation of a people lie in its intelligentsia, in the intellectuals who think honestly, feel, and can work.

—Anton Chekhov, 1921

I saw that most men only care for science so far as they get a living by it, and that they worship even error when it affords them a subsistence.

—Johann Wolfgang von Goethe, 1825

There are very few persons who pursue science with true dignity.

—Humphry Davy, 1830

The degradation of the position of the scientist as an independent worker and thinker to that of a morally irresponsible stooge in a science-factory has proceeded even more rapidly and devastatingly than I had expected.

—Norbert Wiener, 1948

---

[14]As quoted, without a date, by J. F. Maurice in 1814.

Scientists have lost their taste for self-policing and quality control.

—The Economist (editorial), 2013[15]

The history of science swarms with cases of outright fakery and instances of scientists who unconsciously distorted their work by seeing it through lenses of passionately held beliefs.

—Martin Gardner, 1981

Historically, the claim of consensus has been the first refuge of scoundrels. [...] Whenever you hear the consensus of scientists agrees on something or other, reach for your wallet, because you're being had.

—Michael Crichton, 2003

Science [...] commits suicide when it adopts a creed.

—Thomas Henry Huxley, 1885

As an adolescent I aspired to lasting fame, I craved factual certainty, and I thirsted for a meaningful vision of human life—so I became a scientist. This is like becoming an archbishop so you can meet girls.

—Matt Cartmill, 1988

In the sciences the authority of thousands of opinions is not worth as much as one tiny spark of reason in an individual.

—Galileo Galilei, 1612

To be independent of public opinion is the first formal condition or achieving anything great or rational—whether in life or in science.

—Georg Wilhelm Freidrich Hegel, 1821

One of the great commandments of science is, 'Mistrust arguments from authority.' (Scientists, being primates, and thus given to dominance hierarchies, of course do not always follow this commandment.)

—Carl Sagan, 1996

Skepticism is [...] the first step toward truth.

—Denis Diderot, 1746

The path of sound credence is through the thick forest of skepticism.

—George Jean Nathan, 1924

When you know you're right, you don't care what others think. You know sooner or later it will come out in the wash.

—Barbara McClintock, 1983

***

---

[15]In the accompanying detailed article *How Science Goes Wrong*, this conclusion is supported by a set of rather appalling facts and numbers.

The popularization of scientific doctrines is producing as great an alteration in the mental state of society as the material applications of science are effecting in its outward life.

—James Clerk Maxwell, 1871

An alleged scientific discovery has not merit unless it can be explained to a barmaid.

—Ernest Rutherford[16]

A little inaccuracy sometimes saves tons of explanation.

—Saki, 1924

There is no real popularization possible, only vulgarization that in most instances distorts the discoveries beyond recognition.

—Erwin Chargaff, 1973

Science, history and politics are not suited for discussion except by experts.

—Frank Ramsey, 1925

***

The truth is, there is nothing—there is nothing—of the same order as magnitude as the invention of quantum mechanics or of the double helix or of relativity. Just nothing like that has happened in the last few decades.

—Leo Kadanoff[17]

The US must double the amount of research effort searching for new ideas every 13 years to offset the increased difficulty of finding new ideas.

—Nicholas Bloom et al., 2020

It is clear that we cannot go up another two orders of magnitude as we have climbed the last five. If we did, we should have two scientists for every man, woman, child, and dog in the population, and we should spend on them twice as much money as we had. Scientific doomsday is therefore less than a century distant.

—Derek J. de Solla Price, 1963

***

In picking that problem be sure to analyze it carefully to see that it is worth the effort. It takes just as much effort to solve a useless problem as a useful one.

—Charles F. Kettering, 1955

– Most important ideas are not exciting.
– Most exciting ideas are not important.

—Dan Geer, 2007

There is much pleasure to be gained from useless knowledge.

—Bertrand Russell, 1935

---

[16]As quoted, without a date, by G. J. Whitrow in 1973.

[17]As quoted, without a date, by J. Horgan in 1996.

Basic research is like shooting an arrow into the air and, where it lands, painting a target.

—*Homer Burton Adkins*, 1984[18]

While practical benefits often result from pure academic research [...], such benefits are not guaranteed and cannot be predicted; nor need they be seen as the ultimate goal.

—*Abraham Flexner*, 1939

This 'truth for truth's sake' is a delusion of so-called savants.

—*Karl K. Darrow*, 1934

The role of science is auxiliary; it is just a means to the attainment of well-being.

—*Dmitri Mendeleev*[19]

No style of thinking will survive which cannot produce a usable product, when survival is at stake.

—*Thomas Favill Gladwin*, 1970

Pure scientists have by and large been dim-witted about engineers and applied science. [...] Their instinct—perhaps sharpened in this country by the passion to find new snobbism wherever possible, and to invent one if it does not exist—was to take for granted that applied science was an occupation for second-rate minds.

—*C. P. Snow*, 1964

Applied science is just as interesting as pure science, and what is more it's damn sight more difficult.

—*William Bate Hardy*[20]

Fundamental science does things right; engineering does the right things; applied science does the right things right.

—*Anonymous*[21]

The scientist merely explores that which exists, while the engineer creates what has never existed before.

—*Theodore von Kármán*, 1967

Science may amuse and fascinate us, but it is engineering that changes the world. [...] A scientist can discover a star, but he cannot make one. He would have to ask an engineer to make one for him.

—*Isaac Asimov*, 1969

Science clears the fields on which technology can build.

—*Werner Heisenberg*, 1958

---

[18]A more general paraphrase of this quip by Ashleigh Brilliant is even more quoted.

[19]As quoted, without a date, by V. V. Vorontsov in 1979.

[20]As quoted, without a date, by Henry Tizard in 1955.

[21]The expression "doing the right things right" may be met in many texts published well before 2016 when it was taken by Laura Stack for the title of her popular how-to book.

[Engineering] stands on scientific foundations, but there is a big gap between scientific research and the engineering product, which has to be bridged by the art of the engineer.

—Anonymous[22]

***

The job of the engineer is to change the world.

—Robert C. Dean, Jr.[23]

This is not the age of pamphleteers. It is the age of the engineers. The spark-gap is mightier than a pen.

—Lancelot Hogben, 1938

If a country has plenty of creative engineers doing real creative work, it moves forward with times. If it does not, it falls behind, however good all its other people are.

—Meredith Thring, 1964

***

Machine does not isolate man from the great problems of nature but plunges him more deeply into them.

—Antoine de Saint-Exupéry, 1939

Major changes of thought have, in the past, occurred as consequences of technological advances.

—D. S. L. Cardwell, 1973

Nothing tends so much to the advancement of knowledge as the application of a new instrument.

—Humphry Davy, 1812

The science of thermodynamics owed more to the steam engine than the steam engine owed to science.

—Lawrence Joseph Henderson, 1917

Before the 1880s, science played almost no role in advances of technology. For example, James Watt developed the first efficient steam engine long before science established the equivalence between mechanical energy and heat.

—Edward Teller, 2001

In fact, most people—when they speak of science as a good thing—have in mind such technology as has altered the condition of their life.

—J. Robert Oppenheimer, 1959

---

[22]Certain "British engineer" quoted, without a date, by Walter G. Vincenti in 1990.

[23]As quoted, without a date, by Daniel V. DeSimone in 1968.

Technology is the gift of God. [...] It is the mother of civilizations, of arts and sciences.

—*Freeman Dyson*, 2004[24]

Technology is [...] a fusion of nature and the human spirit into a new kind of creation that transcends both.

—*Robert M. Pirsig*, 1974

Technology paces industry. [...] Industry paces economics. [...] And in that manner we come finally to everyday life.

—*Buckminster Fuller*, 1962

Technology has made large populations possible; large populations now make technology indispensable.

—*Joseph Wood Krutch*, 1959

The imperatives of technology and organization, not the images of ideology, are what determine the shape of economics.

—*John Kenneth Galbraith*, 1985

As the world of chips and glass fibers and wireless waves go, so goes the rest of the world.

—*Kevin Kelly*, 1999

***

Any sufficiently advanced technology is indistinguishable from magic.

—*Arthur C. Clarke*, 1973

Civilization advances by extending the number of important operations which we can perform without thinking of them.

—*Alfred North Whitehead*, 1911

Technology [is] a knack of so arranging the world that we don't have to experience it.

—*Max Frisch*, 1957

Thanks to the Interstate Highway System, it is now possible to travel from coast to coast without seeing anything.

—*Charles Kuralt*, 1985

Technology has made improvements in everything, except the weather and people.

—*Evan Esar*, 1968

The development of technology will leave only one problem: the infirmity of human nature.

—*Karl Kraus*, 1976

***

---

[24]Note that this opinion, and the two before it, belong to prominent theoretical physicists, rather than some committed technologists.

Technology [...] is a queer thing. It brings you great gifts with one hand, and it stabs you in the back with the other.

—C. P. Snow, 1971 [25]

For a list of all the ways technology has failed to improve the quality of life, please press three.

—Alice Kahn[26]

We are stuck with technology when what we really want is just stuff that works.

—Douglas Adams, 2002

If it works, it is out of date.

—Stafford Beer, 1981

The technology of mass-production is inherently violent, ecologically damaging, self-defeating in terms of non-renewable resources, and stultifying the human person.

—E. F. Schumacher, 1973

Modern technology
owes ecology
an apology.

—Alan M. Eddison, 1969

Alas! can we ring the bell backward? Can we unlearn the arts that pretend to civilize, and then burn the world? There is a march of science; but who shall beat the drums for its retreat?

—Charles Lamb, 1830

Whatever Nature has in store for mankind, unpleasant as it may be, men must accept, for ignorance is never better than knowledge.

—Enrico Fermi[27]

Since we have no choice but to be swept along by this vast technological surge, we might as well learn to surf.

—Michael E. Soulé, 1989

The question of engineering should be of interest not only to those of us who are engineers, but to the entire public which lives in an engineering world.

—Karl Taylor Compton, 1955

Novels that leave out technology misrepresent life as badly as Victorians misrepresented life by leaving out sex.

—Kurt Vonnegut, 2005

---

[25]Sometimes misattributed to Carrie Snow.

[26]As quoted, without a date, by George Tillman in 1993.

[27]As quoted, without a date, by Laura Fermi in 1954.

We have become a people unable to comprehend the technology we invent.

—Association of American Colleges, 1985

***

Engineering without imagination sinks to a trade.

—Herbert Hoover, 1952

Necessity is the mother of invention.

—English proverb[28]

Invention is the mother of necessity.

—Thorstein Veblen, 1914

You are too late with a development if [...] people demand it before you yourself recognize it.

—Charles F. Kettering, 1934

We cannot predict the future, but futures can be invented.

—Dennis Garbor, 1963[29]

***

To invent, you need a good imagination and a pile of junk.

—Thomas Edison[30]

An inventor is a poet—a true poet.

—Mark Twain, 1870

Engineering, like poetry, is an attempt to approach perfection. And engineers, like poets, are seldom completely satisfied with their creations.

—Henry Petroski, 1955

Life cycle of a typical [project]: unwarranted enthusiasm, uncritical acceptance, growing concern, unmitigated disaster, search for the guilty, punishment of the innocent, and promotion of the uninvolved.[31]

—Ephraim R. McLean, 1972

Engineers like to solve problems. If there are no problems handy, they create their own problems.

—Anonymous

***

---

[28]Possibly based on Plato's line, "Our need will be the real creator".

[29]A popular later paraphrase of this maxim is: "The best way to predict the future is to invent it".

[30]As quoted, without a date, by Ronald C. Arkin in 1998.

[31]Originally describing "a typical electronic data processing system", this semi-joke, frequently in a modified form, is now commonly used to describe the virtually inevitable stages of any large-scale development project.

Any intelligent fool can make things bigger, more complex, and more violent. It takes a touch of genius—and a lot of courage— to move in the opposite direction.

—E. F. Schumacher, 1973

The price of reliability is the pursuit of the utmost simplicity.

—C. A. R. Hoare, 1980

What we decided to leave out is almost as important as what we put in.

—Joshua Schachter, 2006

Perfection [in design] is finally attained not when there is no longer anything to add, but when there is no longer anything to take away.

—Antoine de Saint-Exupéry, 1939

Novelties come from previously unseen association of old material. To create is to recombine.

—François Jacob, 1977

A complex system that works is invariably found to have evolved from a simple system that worked. A complex system designed from scratch never works and cannot be patched to work.[32]

—John Gall, 1975

There's no system foolproof enough to defeat a sufficiently great fool.

—Edward Teller, 1988

Design can be art. Design can be aesthetics. Design is so simple, that's why it is so complicated.

—Paul Rand, 2008

Designing your product for monetization first, and people second will probably leave you with neither.

—Tara Hunt, 2009

***

Nature, to be commanded, must be obeyed.

—Francis Bacon, 1620[33]

[Nature] is always fair, just, and patient. But [it] never overlooks a mistake, or makes a smallest allowance for ignorance.

—Thomas Henry Huxley, 1868

Against the laws of nature, there is no appeal.

—Arthur C. Clarke, 1965

---

[32]The so-called *Gall's Law*.

[33]Possibly stemming from on a much earlier (circa 40s AD) line by Seneca: "He who cannot obey, cannot command".

For a successful technology, reality must take precedence over public relations, for Nature cannot be fooled.

—Richard Feynman, 1988

Necessity gives the law.

—Anonymous[34]

The world stands aside to let anybody pass who knows where he is going.

—David Starr Jordan, 1918

You cannot make pancakes without breaking eggs.

—Spanish proverb

A cat in gloves catches no mice.

—European proverb[35]

The only principle that does not inhibit progress is: everything goes.

—Paul Feyerabend, 1975

Hell, there are no rules around here! We are trying to accomplish something!

—Thomas Edison, 1932

Rules are just helpful guidelines for stupid people who can't make up their own minds.

—Anonymous[36]

Rules are made to be broken.

—Anonymous[37]

If you obey all the rules, you miss all the fun.

—Katharine Hepburn[38]

---

[34]This maxim and its close variant, "Necessity knows no law", are probably based on a longer line by Publilius Syrus: "Necessity gives the law without acknowledging one".

[35]Frequently misattributed to Benjamin Franklin, who merely reproduced this old proverb in his *Poor Richard's Almanack* in 1758.

[36]I have seen this line attributed to some Seth Hoffman, but could not confirm this authorship.

[37]Possibly originated from a longer, more qualifying line by Florence Nightingale (circa 1858).

[38]I know she was in movies rather than in science or engineering, but believe that her observation, made in the late 1930s, is very general.

On Sciences and Technologies

A doctrine of nature can only contain so much science proper as there is in it of applied mathematics.

—*Immanuel Kant*, 1786

Mathematics is the queen of the sciences.

—*Carl Friedrich Gauss*, 1856

The Great Architect of the Universe now begins to appear as a pure mathematician.

—*James Jeans*, 1930

The miracle of the appropriateness of the language of mathematics for the formulation of the laws of physics is a wonderful gift which we neither understand nor deserve.

—*Eugene Wigner*, 1960

Whether or not you ever again use the math you learned in school, the act of having learned the math established a wiring in your brain that didn't exist before, and it's the wiring in your brain that makes you a problem solver.

—*Neil deGrasse Tyson*, 2011

Poor arithmetic will make the bridge fall down just as surely as poor physics, poor metallurgy, or poor logic will.

—*Thomas T. Woodson*, 1966

Bridges would not be safer if only people who knew the proper definition of a real number were allowed to design them.

—*David Mermin*, 1979

The profound study of nature is the most fertile source of mathematical discoveries.

—*Joseph Fourier*, 1822

K. K. Likharev (ed.), *Essential Quotes for Scientists and Engineers*, https://doi.org/10.1007/978-3-030-63332-5_5

It was, no doubt, particularly of [Joseph Fourier's] very disregard for rigor that he was able to take conceptual steps which were inherently impossible to men of more critical genius.

—*Rudolph Ernst Langer*, 1947

Nothing is more repellent to normal human beings than the clinical succession of definitions, axioms, and theorems generated by the labors of pure mathematicians.

—*J. M. Ziman*, 1969

Mathematics may be defined as the subject in which we never know what we are talking about, nor whether what we are saying is true.

—*Bertrand Russell*, 1901

***

As far as the laws of mathematics refer to reality, they are not certain; and as far as they are certain, they do not refer to reality.

—*Albert Einstein*, 1920

If scientific reasoning were limited to the logical processes of arithmetic, we should not get very far in our understanding of the physical world.

—*Vannevar Bush*, 1945

[In physics,] precise logical thinking, typical for mathematicians, inhibits suggestion of new ideas, because it arrests imagination.

—*Pyotr Kapitsa*, 1973

It is really quite impossible to say anything with absolute precision, unless that thing is so abstracted from the real world as to not represent any real thing.

—*Richard Feynman*, 1965

In physics, your solution should convince a reasonable person. In math, you have to convince a person who's trying to make trouble. Ultimately, in physics, you're hoping to convince Nature. And I've found Nature to be pretty reasonable.

—*Frank Wilczek*, 2009

***

Oh, physics! That's too much difficult for physicists.

—*David Hilbert*, 1925

All science is either physics or stamp collecting.

—*Ernest Rutherford*[1]

---

[1] As quoted, without a date, by J. B. Birks in 1962.

Nature and Nature's laws lay hid in night:
God said, *Let Newton be*! and all was light.

—Alexander Pope, 1730

It did not last: the Devil howling 'Ho!
Let Einstein be!' restored the status quo.

—J. C. Squire, 1926

Because of the quantum nature of the problem, one cannot say that the present stage of knowledge is exactly equal to zero.

—Harry J. Lipkin, 1956[2]

Schrödinger called his cat and said,
'You can be both alive and dead,
For a linear combination of states
Postulates two simultaneous fates'.[3]

—M. Kocher, 1978

Today we no longer ask what really goes on in an atom, we ask what is likely to be to be observed – and with what likelihood.

—Otto Robert Frisch, 1979

A professor of theoretical physics always has to be told what to look for. He just uses his knowledge to explain the observations of the experimenters.

—Richard Feynman, 1988

I am convinced that [God] does not play dice.

—Albert Einstein, 1926

Bohr could only counter with 'Nor it is our business to prescribe God how He should run the world'.

—Werner Heisenberg, 1969

Bohr was inconsistent, unclear, willfully obscure - and right. Einstein was consistent, clear, down-to-earth - and wrong.

—John Bell, 2010[4]

Even if we understand all the laws of physics, then exploring their consequences in the everyday world, where complex structures can exist, is a far more daunting task, and that's an inexhaustible one I'm sure.

—Martin John Rees, 1988

---

[2]From his article *Theoretical Zipperdynamics*, signed by pen name Harry J. Zipkin, and published in the famous *Journal of Irreproducible Results*. (Lipkin was one of its founders.).

[3]To the so-called *Schrödinger Cat paradox* of quantum mechanics.

[4]J. Bell's opinion about the famous Einstein-Bohr debate (see the previous two quotes) is important, because he is widely credited for the last significant theoretical contribution to this topic (now called the *local reality problem*), made in the 1960s – even if some physicists still deem the problem not fully solved.

A reductionist philosophy, arbitrarily proclaiming that the growth of understanding must go only in one direction, makes no scientific sense.

—*Freeman Dyson*, 1995

***

Chemistry stands at the pivot of science. On the one hand it deals with biology and provides explanations for the processes of life. On the other hand it mingles with physics and finds explanations for chemical phenomena in the fundamental processes and particles of the universe. Chemistry links the familiar with the fundamental.

—*Peter Atkins*, 1987

The chemists are a strange class of mortals, impelled by an almost insane impulse to seek their pleasure amid smoke and vapor, soot and flame, poisons and poverty. Yet among all these evils I seem to live so sweetly that I may die if I were to change places with the Persian king.

—*Johann Joachim Becher*, 1667

A tidy laboratory means a lazy chemist.

—*Jöns Jacob Berzelius*, 1812

A chemist who does not know mathematics is seriously handicapped.

—*Irving Langmuir*[5]

A chemist who is not a physicist is nothing at all.

—*Robert Bunsen*[6]

Chemistry is the dirty part of physics.

—*Johann Philipp Reis*[7]

Chemistry has been named by the physicist as the messy part of physics, but that is no reason why physicists should be permitted to make a mess of chemistry when they invade it.

—*Frederick Soddy*, 1946

Physicists use excellent methods to study poor materials, chemists use poor methods to study excellent materials, and physical chemists use poor methods to study poor materials.

—*Hans Heinrich Landolt*[8]

Between physics and chemistry, it is hard to know who should study what molecule.

—*Philip Morrison*, 1995

***

---

[5]As quoted, without a date, by Albert Rosenfeld in 1962.

[6]As quoted, without a date, by J. R. Partington in 1961.

[7]As quoted, without a date, by R. Oesper in 1975.

[8]To estimate this semi-joke, note that the author (of the Landolt-Börnstein database fame) was one of the pioneers of physical chemistry.

Geography is just physics slowed down, with a couple of trees stuck in it.

—*Terry Pratchett, 2008*

As a young man, my fondest dream was to become a geographer. However... I thought deeply about the matter and concluded that it was far too difficult a subject. With some reluctance, I then turned to physics as a substitute.

—*Duane Francis Marble*[9]

Geology is the music of the earth.

—*Hans Cloos, 1954*

We learn geology the morning after the earthquake.

—*Ralph Waldo Emerson, 1860*

A geologist is a fault finder.

—*Bob Phillips, 2010*

Continental drift was guilty until proven innocent.

—*David M. Raup, 1986*

***

Statistics: The only science that enables different experts using the same figures to draw different conclusions.

—*Evan Esar, 1943*

Politicians use statistics in the same way that a drunk uses lampposts – for support rather than illumination.

—*Andrew Lang, 1910*

Smoking is one of the leading causes of statistics.

—*Fletcher Knebel, 1965*

Average a left-hander with a right-hander, and what do you get?[10]

—*Don Norman, 1988*

91.7% of all statistics is done on spot, including this number.

—*Anonymous*

There are three kind of lies: lies, damned lies, and statistics.

—*Anonymous*[11]

---

[9]Posted (in the 1960s) on his office door, signed *Albert Einstein*, as a joke, provoking quite a few misattributions.

[10]Cf. the Ehrenfest theorem in quantum mechanics, infamous for its lack of insight into many key quantum effects including entanglement.

[11]Misattributed to many, including Benjamin Disraeli and Mark Twain.

Fate laughs at probabilities.

—Edward Bulwer-Lytton, 1832

***

Computer science is not about machines, in the same way that astronomy is not about telescopes.

—Michael R. Fellows, 1991[12]

Computer science is neither mathematics nor electrical engineering.

—Alan Perlis, 1968

Computer science […] is not actually a science.

—Richard Feynman, 1970

Any discipline with the word 'science' in its name, such as social sciences, creation science, or computer science, is not a science.

—Anonymous[13]

Perhaps the central problem we face in all of computer science is how we are to get to the situation where we build on top of the work of others rather than redoing so much of it in a trivially different way.

—Richard Hamming, 1968

Software engineering is that part of computer science which is too difficult for computer scientists.

—Friedrich L. Bauer, 1971

***

The beauty of a living thing is not the atoms that go into it, but the way the atoms are put together.

—Carl Sagan, 1990

Biology […] is the science that stands at the center of all science.

—George Gaylord Simpson, 1964

Seen in the light of evolution, biology is, perhaps, intellectually the most satisfying and inspiring science.

—Theodosius Dobzhansky, 1972

Evolution advances, not by a priori design, but by the selection of what works best out of whatever choices offer. We are products of editing, rather than of authorship.

—George Wald, 1957

---

[12]Possibly based on an earlier (circa 1986) but longer line by Hal Abelson. A close paraphrase of this maxim is frequently misattributed to Edsger W. Dijkstra.

[13]This popular remark is attributed to various scientists, most plausibly to Hal Abelson and Frank Harary, but I could not find a reliable confirmation of either authorship.

While Occam's Razor[14] is a useful tool in physical sciences, it can be a very dangerous implement in biology. It is thus very rash to use simplicity and elegance in biological research.

—Francis Crick, 1988

Biology belongs to one of the surprising sciences, where each rule must always be supplemented with several exceptions (except this rule, of course).

—Claus Emmeche, 1994

Physics was the first of the natural sciences to become fully modern and highly mathematical. Chemistry followed in the wake of physics, but biology, the retarded child, lagged far behind.

—Michael Crichton, 1969

Biology has become a mature science as it has become precise and predictable.

—Philip Handler, 1970

What is truly revolutionary about molecular biology in the post-Watson-Crick era is that it became digital.

—Richard Dawkins, 1995

***

Aristotle was famous for knowing everything. He taught that the brain exists merely to cool the blood and is not involved in the process of thinking. This is true only of certain persons.

—Will Cuppy, 1950

The brain is the most complex thing we have yet discovered in the universe.

—James Watson, 1992

A typical neuron makes about ten thousand connections to neighboring neurons. Given the billions of neurons, there are as many connections in a single cubic centimeter of brain tissue as there are stars in the Milky Way galaxy.

—David Eagleman, 2011

The stars may be large, but they cannot think or love; and these are qualities which impress me far more than the size does. I take no credit for weighing seventeen stone.

—Frank Ramsey, 1931

The most underdeveloped territory in the world is under our scalps.

—Dorothy M. Carl, 1972[15]

---

[14]"Plurality [of reasons] should not be postulated without necessity" (Lat. "*Pluralitas non est ponenda sine necessitate*") by William of Ockham. This principle is widely misquoted as "Entities should not be multiplied unnecessarily".

[15]Unfortunately, I could not identify this author.

Brains cause technology, society, art, science, soap operas, sin. A remarkable set of effects for such a small chunk of coagulated atoms.

—*Collin McGinn*, 1999

I bet the human brain is a kludge.

—*Marvin Minsky*[16]

If the human mind was simple enough to understand, we'd be too simple to understand it.

—*Emerson M. Pugh*, ~1938[17]

The human brain is the last, and greatest scientific frontier.[18]

—*Joel L. Davis*, 1997

***

[Medicine:] the art of amusing the patient while nature cures the disease.

—*Ben Jonson*[19]

God cures, and the doctor takes the fee.

—*European proverb*

The blunders of physicians are covered by the earth.

—*Spanish proverb*

Medicine is a science of uncertainty and the art of probability.

—*William Osler*, 1950

It is my duty to heal the sick, not to enrich the apothecaries.

—*Paracelsus*[20]

One of the first duties of the physician is to educate the masses not to take medicine.

—*William Osler*, 1961

A drug is a substance which, if injected into a rabbit, produces a [scientific] paper.

—*Otto Loewi*[21]

---

[16]As quoted, without a date, by David K. Mellinger in 1991.

[17]First quoted by his son, George W. Pugh, in 1977.

[18]Cf. the expression "Space, the final frontier" used in the *Star Trek* movies.

[19]As quoted, without a date, in an editorial of *The Veterinarian* in 1851; frequently misattributed to Voltaire.

[20]As quoted, without a date, by many, possibly starting with Joseph Ennemoser and Mary Botham Howitt in 1854.

[21]As quoted by A. Szent-Gyorgyi in 1976; paraphrased, after O. Loewi's death in 1961, by others —mostly without a reference.

The best doctors are Dr. Diet, Dr. Quiet, and Dr. Merryman.

—English proverb[22]

***

A vigorous five-mile walk will do more good for an unhappy but otherwise healthy adult than all the medicine and psychology in the world.

—Paul Dudley White[23]

What is true [in psychology] is alas not new, the new is not true.

—Hermann Ebbinghaus, 1873

[Modern psychology] appears to be the sickly offspring of average common sense.

—Ludwig Klages, 1929

The psychiatrist unfailingly recognizes the madman by his excited behavior on being incarcerated.

—Karl Kraus[24]

Anyone who goes to a psychiatrist should have his head examined.

—Samuel Goldwyn[25]

Psychoanalysis is a permanent fad.

—Peter De Vries, 1973

***

[The economist] must be suspicious of any direct light that the past is said to throw on the problems of the present.

—Alfred Marshall, 1925

The only function of economic forecasting is to make astrology look respectable.

—Ezra Solomon, 1984[26]

An economist is an expert who will know tomorrow why the things he predicted yesterday didn't happen today.

—Laurence J. Peter, 1977

If all economists were laid end to end, they would not reach a conclusion.

—Anonymous[27]

---

[22]Essentially, a succinct version of an earlier (circa 1449) line by John Lydgate.

[23]As quoted, without a date, by William Fitzgibbon in 1972.

[24]As quoted, without a date, by many, starting perhaps with Thomas Szasz in 1976.

[25]As attributed in 1948; it is suspected that Lillian Hellman was the real author.

[26]Frequently misattributed to John Kenneth Galbraith.

[27]Frequently misattributed to George Bernard Shaw.

Economics is based on the assumption that people have reasonably simple objectives and choose the correct means to achieve them. Both assumptions are false – but useful.

—*David Friedman*, 1996

In economics, hope and faith coexist with great scientific pretension.

—*John Kenneth Galbraith*, 1970

***

There is nothing so absurd but some philosopher has said it.

—*Cicero*, 44 BC

There are more things in heaven and earth, Horatio,
Than are dreamt of in your philosophy.

—*William Shakespeare*, 1603

Philosophy is nothing but discretion.

—*John Selden*, 1689

Philosophy […] consists chiefly in suggesting unintelligible answers to insoluble problems.

—*Henry Adams*, 1907

Philosophy, *n*. A route of many roads leading from nowhere to nothing.

—*Ambrose Bierce, 1911*

A philosopher is a blind man in a dark room looking for a black cat that isn't there. A theologian is the man who finds it.

—*Anonymous*[28]

Philosophers no longer write for the intelligent, only for their fellow professionals.

—*Jacques Barzun*, 1989

Philosophy consists very largely of one philosopher arguing that all others are jackasses. He usually proves it, and I should add that he also usually proves that he is one himself.

—*H. L. Mencken*, 1956

Philosophers, incidentally, say a great deal about what is absolutely necessary for science, and it is always, so far as one can see, rather naive, and probably wrong.

—*Richard Feynman*, 1963[29]

---

[28]Published as an anonymous joke by H. L. Mencken in 1942. It might be based on a similar statement (about the term "equity") by Charles Bowen, quoted in 1911 by John Alderson Foote.

[29]I could not find a reliable confirmation of the authorship of another famous aphorism on this subject, sometimes attributed to R. Feynman and sometimes to S. Weinberg: "The philosophy of science is as useful to scientists as ornithology is to birds." (The same is frequently said about the value of aesthetics for arts).

It is far safer and wiser that the physicist remain on the solid ground of theoretical physics itself and eschew the shifting sands of philosophic extrapolations.

—*Louis de Broglie*, 1962

People complain that our generation has no philosophers. They are wrong. They now sit in another faculty. Their names are Max Planck and Albert Einstein.

—*Adolf Harnack*, 1911

***

The most ominous conflict of our time is the difference of opinion, of outlook, between men of letters, historians, philosophers, the so-called humanists, on the one side and scientists on the other. The gap cannot but increase because of the intolerance of both and the fact that science is growing by leaps and bounds.

—*George Sarton*, 1930

The intellectual life of the whole western society is increasingly being split into two polar groups. [...] At one pole we have the literary intellectuals [...], at the other scientists, and as the most representative, the physical scientists. Between the two a gulf of mutual incomprehension.

—*C. P. Snow*, 1959

The backwardness of the economic and social sciences with respect to the sciences of matter is one of the causes of the present human calamities through lack of foresight.

—*Jean Fourastié*, 1949

One of the differences between the natural and the social sciences is that in the natural sciences, each succeeding generation stands on the shoulders of those that have gone before, while in the social sciences, each generation steps in the faces of its predecessors.[30]

—*David Zeaman*, 1959

Science is one of the very few human activities – perhaps the only one – in which the errors are systematically criticized and fairly often, corrected.

—*Karl Popper*, 1963

In science it often happens that scientists say, 'You know that's a really good argument; my position is mistaken,' and then they would actually change their minds and you never hear that old view from them again. [...] I cannot recall the last time something like that happened in politics or religion.

—*Carl Sagan*, 1987

There is a noticeable general difference between the sciences and mathematics on the one hand, and the humanities and social sciences on the other. [...] You can lie or distort the story of the French Revolution as long as you like, and nothing will happen. Propose a false theory in chemistry, and it'll be refuted tomorrow.

—*Noam Chomsky*, 1992

---

[30] As a member of the natural science community, I am grateful for author's high opinion of it, but cannot help noticing that some of its dwarfish members use their elevated positions to relieve themselves on the heads of their predecessors.

The specialized languages of [social sciences] serve virtually no other purpose than to conceal valuation behind an ostensibly scientific and therefore nonvaluational semantic screen.

—*Thomas Szasz*, 1990

Praise up the humanities, my boy, that'll make them think that you're broad-minded!

—*Winston Churchill*, 1946[31]

***

Computing is not about computers anymore. It is about living.

—*Nicolas Negroponte*, 2015

[The computer] seems to me to be an Old Testament god with a lot of rules and no mercy.

—*Joseph Campbell*, 1988

On two occasions I have been asked, by members of Parliament, 'Pray Mr. Babbage, if you put into the machine the wrong figures, will the right answers come out?' [...] I am not able rightly to apprehend the kind of confusion of ideas that could provoke such a question.

—*Charles Babbage*, 1864[32]

Garbage in, garbage out.

—*Anonymous*[33]

I spend almost as much time figuring out what's wrong with my computer as I do actually using it.

—*Clifford Stoll*, 1996

That's the thing about people who think they hate computers... What they really hate are lousy programmers.

—*Larry Niven*, 1982

Software engineering has accepted as its charter 'How to program if you cannot.'

—*Edsger W. Dijkstra*, 1988

Most software today is very much like an Egyptian pyramid with millions of bricks piled on top of each other, with no structural integrity, but done just by brute force and thousands of slaves.

—*Alan Kay*, 2005

---

[31]As quoted in 1996 by R. V. Jones, to whom this advice was given.

[32]This was about his *Difference Engine No. 1*—essentially the first computer.

[33]This elegant twist of the term *First-In, First-Out* (FIFO) *register* has turned into a major principle of computer applications. Created in the mid-1950s, it is typically attributed to either Stephen Wilfred ("Wilf") Hey or E. E. Blanche, but I could not reliably confirm these authorships and even identify these persons.

If the automobile had followed the same development cycle as the computer, a Rolls-Royce would today cost $100, get a million miles per gallon, and explode once a year, killing everyone inside.

—Robert X. Cringely, 1969[34]

If builders built houses the way programmers build programs, the first woodpecker to come along would destroy civilization.

—Gerald Marvin Weinberg[35]

A refund for defective software might be nice, except it would bankrupt the entire software industry in the first year.

—Andrew S. Tannenbaum, 2003

It's scary to think that the infrastructure of the industrialized world is increasingly based on software like this.[36]

—Anonymous[37]

It's the basic fact that all programming languages suck.

—Larry Wall, 2006[38]

The purpose of most computer languages is to lengthen your resume by a word and a comma.

—Anonymous[39]

There is more disputing about the shell then the kernel.

—German proverb[40]

The most likely way for the world to be destroyed, most experts agree, is by accident. That's where we come in; we're computer professionals. We cause accidents.

—Nathaniel Borenstein, 1991

***

---

[34]With the ongoing development of autonomous (computer-driven) cars, the last goal seems within reach.

[35]As quoted, without a date, by Murali Chemuturi in 2010.

[36]This was apparently written in 1992 about *AutoCAD* rev. 12. In comparison with that program, most software packages I have to use nowadays are hardware-resource-wasting junk.

[37]I have seen this statement attributed to some Stephen Wolfe, but I was unable to confirm this authorship.

[38]Note that L. Wall has himself created one of the languages (*Perl*), and did not exclude it from this rule.

[39]Sometimes also attributed to the same Larry Wall, but I could not find a reliable confirmation of his authorship.

[40]I find it remarkable how well some old wisdoms work in new contexts – in this case, of the computer operation system development.

By this Contrivance, the most ignorant Person at a reasonable Charge, and with a little body Labor, may write Books in Philosophy, Poetry, Politics, Law, Mathematics and Theology, without the least Assistance from Genius or Study.

—Jonathan Swift, 1724[41]

There is no security [...] against the ultimate development of mechanical consciousness.

—Samuel Butler, 1872

I propose to consider the question, 'Can machines think?'.

—Alan Turing, 1950

The real problem is not whether machines think but whether men do.

—B. F. Skinner, 1969

A computer would deserve to be called intelligent if it could deceive a human into believing that it was human.[42]

—Alan Turing, 1950

Any artificial intelligence smart enough to pass a Turing test is smart enough to know to fail it.

—Ian McDonald, 2006

We're rapidly creating an extraordinary silicon-based Petri dish for evolution of intelligence. By the year 2025 [...] we're likely to have computers whose raw processing power exceeds that of the human brain. Also, we're likely to have more computers than people.

—J. Doyne Farmer, 1995[43]

We cannot prevent the *Singularity*,[44] that is coming as an inevitable consequence of the humans' natural competitiveness and the possibilities inherent to technology.

—Vernor Vinge, 1993

The development of full artificial intelligence could spell the end of the human race. We cannot quite know what will happen if a machine exceeds our own intelligence, so we can't know if we'll be infinitely helped by it, or ignored by it and sidelined, or conceivably destroyed by it.

—Stephen Hawking, 2014

---

[41]Ridiculing the artificial intelligence hopes of his time.

[42]The original suggestion of what is now called the *Turing Test*.

[43]Note that both these thresholds were reached at least 5 years before the supposed date.

[44]In his earlier (1987) book, Vinge defined the *Singularity* as the hypothetical point of "creation of intelligences greater than our own". (Several earlier authors, starting at least from John von Neumann, are credited for similar notions, under different names. The *Singularity* term has received an additional strong promotion from a 2005 book by Ray Kurzweil.).

[The developers of conventional artificial intelligence] have to admit that deep learning is doing amazing things, and they want to use [it] as a kind of low-level servant to provide them with what they need to make their symbolic reasoning work.

—Geoffrey Hinton, 2018[45]

---

[45]Hinton is one of the pioneers of the current revolution in deep learning (DL)—a machine learning technique frequently oversold as artificial intelligence (AI). Actually, the DL is currently limited to pattern classification, not explicitly addressing general AI tasks—the main goal of the conventional, symbolic approaches to AI, snubbed by Hinton. (To be fair, these approaches, indeed, are not showing a nearly fast progress, while their proponents, in turn, frequently snub neural-network techniques such as DL as primitive *connectionist models*.) More generally, I am sorry for finding too few suitable quotes on these topics, which I believe are of paramount importance for our civilization.

On Education, Reading, and Knowledge

Education is the transmission of civilization.

—Will and Ariel Durant, 1968

Human history becomes more and more a race between education and catastrophe.

—H. G. Wells, 1920

Learning is not compulsory; it's voluntary. [...] But to survive, we must learn.[1]

—W. Edwards Deming, 1995

The brighter you are, the more you have to learn.

—Don Herold, 1926

***

Good teaching is forever, and the teacher is immortal.

—Jesse Stuart, 1958

Teaching is not a lost art, but the regard for it is a lost tradition.

—Jacques Barzun, 1945

America believes in education: the average professor earns more money in a year than a professional athlete earns in a whole week.

—Evan Esar, 1943[2]

Good teaching is one-fourth preparation and three-fourths theater.

—Gail Godwin, 1974

***

---

[1]Frequently paraphrased as "Learning is not compulsory; neither is survival".

[2]Note the date. The examples I see around nowadays indicate that this optimistic comparison is no longer valid.

K. K. Likharev (ed.), *Essential Quotes for Scientists and Engineers*,
https://doi.org/10.1007/978-3-030-63332-5_6

Men learn while they teach.

—*Seneca*, 64 AD

You teach best what you most need to learn.

—*Richard Bach*, 1977

He who can, does. He who cannot, teaches.

—*George Bernard Shaw*, 1903

He who cannot even teach how to do, teaches how to teach.

—*Anonymous*[3]

Any teacher that can be replaced by a machine, should be.

—*Arthur C. Clarke*, 1980

***

For every person who wants to teach there are approximately thirty people who don't want to learn - much.

—*W. C. Sellar and R. J. Yeatman*, 1932

Human beings, who are almost unique in having the ability to learn from the experience of others, are also remarkable for their apparent disinclination to do so.

—*Douglas Adams* and *Mark Carwardine*, 1990

Educate gradually, avoiding bloodshed when possible.

—*Mikhail Saltykov-Shchedrin*, 1870

Studying without a liking for it spoils the memory, and it retains nothing it takes in.

—*Leonardo da Vinci*[4]

You can take a horse to the water, but you can't make him drink.

—*European proverb*

Give neither salt nor advice, until you are asked for it.

—*European proverb*

He who is ashamed of asking, is ashamed of learning.

—*Danish proverb*

Teachers open the door, you enter yourself.

—*Chinese proverb*

---

[3]I saw this natural extension of the Shaw's maxim mentioned in print as early as in 1955, already as a known joke. (It was used, in particular, to describe the former USSR Academy of Pedagogical Sciences.)

[4]Undated entry in one of his notebooks, as translated by J. P. Richter in 1883.

The best teacher is the one who [...] inspires his listener with the wish to teach himself.

—Edward Bulwer-Lytton, 1873

What we badly need is social approval of learning and social rewards for learning.

—Isaac Asimov, 1980

***

You don't understand anything until you learn it more than one way.

—Marvin Minsky, 2005

I hear and I forget.
I see and I remember.
I do and I understand.

—Xun Kuang[5]

One must learn by doing the thing; for though you think you may know it, you have no certainty until you try.

—Sophocles[6]

I profess to learn and to teach anatomy not from books but from dissections, not from tenets of Philosophers but from the fabric of Nature.

—William Harvey, 1628

An ounce of practice is worth a pound of precept.

—English proverb

***

To study and not think is waste. To think and not study is dangerous.

—Confucius, fifth century BC

In youth we learn; in age we understand.

—Marie von Ebner-Eschenbach, 1893

Sense comes with age.

—Spanish proverb

If youth only knew; if age only could.

—Henri Estienne, 1594

Wisdom comes, jump goes.

—Māris Liepa[7]

---

[5]Sometimes misattributed to Confucius, whose follower Xun Kuang was.

[6]From his play *Women of Trachis*, whose first English translation was apparently published only in 1904.

[7]As popularly attributed, without a date, to this ballet dancer famous for his jumps.

The older I grow the more I distrust the familiar doctrine that age brings wisdom.

—*H. L. Mencken, 1922*

The man who is too old to learn was probably always too old to learn.

—*Henry S. Haskins, 1940*

***

Many much-learned men have no intelligence.

—*Democritus*[8]

Intelligence [....] enables a man to get along without education. Education [...] enables a man to get along without the use of intelligence.

—*Albert Edward Wiggam, 1924*

The central task of education is to implant a will and facility for learning; it should produce not learned but learning people.

—*Eric Hoffer, 1973*

An education [is] being able to differentiate between what you do know and what you don't.

—*William Feather, 1956*[9]

Education, *n.* That which discloses to the wise and disguises from the foolish their lack of understanding.

—*Ambrose Bierce, 1906*

Education is what most receive, many pass on, and few possess.

—*Karl Kraus, 1976*

Education is the inculcation of the incomprehensible into the ignorant by the incompetent.

—*Josiah Stamp, 1933*[10]

Education is what survives when what had been learned has been forgotten.

—*B. F. Skinner, 1964*[11]

Education is a method by which one acquires a higher grade of prejudices.

—*Laurence J. Peter, 1977*

Without education, we are in a horrible and deadly danger of taking educated people seriously.

—*G. K. Chesterton, 1907*

---

[8]As quoted, without the date, in several reputable sources, starting not later than 1907.

[9]Frequently misattributed to Anatole France.

[10]Frequently misattributed to John Maynard Keynes.

[11]Similar and earlier statements, but about culture rather than education, are attributed to Édouard Herriot and José Ortega y Gasset.

This is one of those views which are so absolutely absurd that only very learned men could possibly adopt them.[12]

—Bertrand Russell, 1959

I prefer the company of peasants because they have not been educated sufficiently to reason incorrectly.

—Michel de Montaigne[13]

It is better to deal with a whole fool than half a fool.

—German proverb

A little Learning is a dang'rous Thing.
Drink deep, or taste not the Pierian Spring.

—Alexander Pope, 1709

If a little knowledge is dangerous, where is the man who has so much as to be out of danger?

—Thomas Henry Huxley, 1907

***

School is learning things you don't want to know, surrounded by people you wish you didn't know, while working toward a future you don't know will ever come.

—Dave Kellett, 2011

A child only educated at school is an uneducated child.

—George Santayana, 1935

John has been to school.
To learn to be a fool.

—European proverb

I have never let schooling interfere with my education.

—Mark Twain, 1869

It is, in fact, nothing short of a miracle that the modern methods of instruction have not entirely strangled the holy curiosity of inquiry.

—Albert Einstein, 1949

Every kid starts out as a natural-born scientist, and then we beat it out of them. A few trickle down the system with their wonder and enthusiasm for science intact.

—Carl Sagan, 1996

***

---

[12]About a certain pronouncement of some, in Russell's words, "modern philosophers".

[13]As quoted, without a date, in several reliable collections, including *Wikiquotes*.

College isn't the place to go for any ideas.

—Helen Keller, 1916

[Colleges and universities] have become intellectually monochrome purveyors of groupthink.

—George Will, 2020

A fool's brain digests philosophy into folly, science into superstition, and art into pedantry. Hence university education.

—George Bernard Shaw, 1903

A university develops all abilities, including the foolishness.

—Anton Chekhov, 1921

Strange as it may seem, no amount of learning can cure stupidity, and formal education positively fortifies it.

—Stephen Vizinczey, 1975

A university is what a college becomes when the faculty loses interest in students.

—John Ciardi, 1966

Nature is not organized in the way universities are.

—Russell Lincoln Ackoff, 1970

There is a tolerably good agreement about what a university is not. It is not a place of professional education.

—John Stuart Mill, 1867[14]

A young man passes from our public schools to the universities, ignorant almost of the elements of every branch of useful knowledge.

—Charles Babbage, 1830

Academe, *n.* An ancient school where morality and philosophy were taught.
Academy, *n.* [from Academe] A modern school where football is taught.

—Ambrose Bierce, 1906

Academia is to knowledge what prostitution is to love: close enough on the surface but […] not exactly the same thing.

—Nassim Nicholas Taleb, 2010

Nothing in education is so astonishing as the amount of ignorance it accumulates in the form of inert facts.

—Henry Adams, 1907

Bachelor's degrees make pretty good placemats if you get 'em laminated.

—Jeph Jacques, 2005

---

[14]From his inaugural address as the rector of the University of St. Andrews.

The average Ph.D. thesis is nothing but a transference of bones from one graveyard to another.

—J. Frank Dobie, 1945

Bob Wilson admitted to himself that a Ph.D. and an appointment as an instructor was not his idea of existence. Still, it beat working for a living.

—Robert A. Heinlein, 2005

***

[What we become] depends on what we read, after all manner of Professors have done their best for us. [...] The true University of these days is a Collection of Books.

—Thomas Carlyle, 1840

The reading of all good books is indeed like a conversation with the noblest men of past centuries.

—René Descartes, 1637

A book is a gift you can open again and again.

—Ron Silliman, 1988[15]

Where books are burned, in the end people will be burned too.

—Heinrich Heine, 1823

An ordinary man can in the ordinary course [...] surround himself with [two thousand books] and thenceforward have at least one place in the world in which it is possible to be happy.

—Augustine Birrell, 1895

A couple of months in the laboratory can frequently save a couple of hours in the library.

—Frank Westheimer, 1979

Education [...] has produced a vast population able to read but unable to distinguish what is worth reading.

—G. M. Trevelyan, 1942

Be as careful of the books you read, as of the company you keep; for your habits and character will be as much influenced by the former as by the latter.

—Edwin Paxton Hood, 1851

We must be careful what we read, and not, like the sailors of Ulysses, take bags of wind for sacks of treasure.

—John Lubbock, 1901

Few books today are forgivable.

—R. D. Laing, 1967

---

[15]Sometimes misattributed to Garrison Keillor.

The reason why so few good books are written is that so few people who can write know anything.

—*Walter Bagehot*, 1853

There are many people who read simply to prevent themselves from thinking.

—*Georg Christoph Lichtenberg*[16]

Reading, after a certain age, diverts the mind too much from its creative pursuits. Any man who reads too much and uses his own brain too little falls into lazy habits of thinking.

—*Albert Einstein*, 1929

Books are good enough in their own way, but they are a mighty bloodless substitute for life.

—*Robert Louis Stevenson*, 1877

***

Knowledge itself is power.[17]

—*Francis Bacon*, 1597

An investment in knowledge always pays the best interest.

—*Anonymous*[18]

All our science, measured against reality, is primitive and childlike – and yet it is the most precious thing we have.

—*Albert Einstein*, 1951

We must know. We will know.

—*David Hilbert*, 1930

***

Knowledge without conscience is but the ruin of the soul.

—*François Rabelais*, 1532[19]

Integrity without knowledge is weak and generally useless, and knowledge without integrity is dangerous and dreadful.

—*Samuel Johnson*, 1759

To educate a man in mind and not in morals is to educate a menace to society.

—*Theodore Roosevelt*[20]

Man armed with science is like a baby with a box of matches.

—*J. B. S. Haldane*, 1924

---

[16]From his *Notebook G* (1779–1783).

[17]This original statement is frequently shortened to just "Knowledge is power".

[18]Broadly attributed to Benjamin Franklin, but not found in Yale's *Franklin Papers*.

[19]Often misattributed to Michel de Montaigne.

[20]As quoted by many, starting not later than 1931.

Our hearts should know the world of reason, and reason must be guided by an informed heart.

—Bruno Bettelheim, 1990

***

There is only one good, knowledge, and one evil, ignorance.

—Socrates[21]

Ignorance is not innocence but sin.

—Robert Browning, 1875

Nothing in all the world is more dangerous than sincere ignorance and conscientious stupidity.

—Martin Luther King Jr., 1963

He who is not aware of his ignorance will only be misled by his knowledge.

—Richard Whatley, 1856

To be proud of learning is the greatest ignorance in the world.

—Jeremy Taylor[22]

The greatest obstacle to [discovery is] not ignorance, but the illusion of knowledge.

—Daniel J. Boorstin, 1983

Very few people, the most remarkable ones, can honestly utter: 'I do not know'.

—Dmitry Pisarev[23]

To be conscious that you are ignorant is a great step to knowledge.

—Benjamin Disraeli, 1845

***

Have the courage to be ignorant of a great number of things, in order to avoid the calamity of being ignorant of everything.

—Sydney Smith, 1849

There are many things of which a wise man might wish to be ignorant.

—Ralph Waldo Emerson, 1883

In research a certain amount of intelligent ignorance is essential to progress.

—Charles F. Kettering, 1957

Man can act only because he can ignore.

—Paul Valéry, 1921

---

[21]As quoted by Diogenes Laërtius in the third century AD.

[22]First published, apparently, only in 1822.

[23]As quoted, without a date, by V. Vorontsov in 1979.

The whole secret of life is to be interested in one thing profoundly and in a thousand things well.

—Horace Walpole[24]

***

Science is the belief in the ignorance of experts.

—Richard Feynman, 1969

Amateurs built the Ark. Professionals built the Titanic.

—Richard J. Needham, 1979[25]

Always listen to experts. They'll tell you what can't be done and why. Then do it.

—Robert A. Heinlein, 1973

If the world should blow itself up, the last audible voice would be that of an expert saying it can't be done.

—Peter Ustinov, 2004

For every expert there is an equal and opposite expert.

—Anonymous[26]

An expert is a person who avoids small error as he sweeps on to the grand fallacy.

—Benjamin Stolberg[27]

An expert is someone licensed to do things he cannot do.

—Erwin Chargaff, 1980

An expert in any field is a person who knows enough about what's really going on to be scared.

—P. J. Plauger, 1983

The price one pays for pursuing any profession or calling is an intimate knowledge of its ugly side.

—James Baldwin, 1961

***

You live and learn. At any rate, you live.

—Douglas Adams, 2000

---

[24]As quoted in *The Christian Leader* in 1934.

[25]Paraphrased later by many, including Elizabeth May (in 2014) and Ben Carson (in 2015).

[26]Frequently attributed to Arthur C. Clarke, but he quotes this *4th Law* as a known joke.

[27]As quoted, without a date, in several generally reliable collections, starting not later than 1952.

On the Wise and Otherwise, Sanity and Madness

Dare to be wise.

—Horace, $\sim$21 BC

Wise men learn by other men's mistakes, fools by their own.[1]

—European proverb

A wise man gets more use from his enemies than a fool from his friends.

—Baltasar Gracián, 1647[2]

There is no fool so great a fool as a knowing fool. But to know how to use knowledge is to have wisdom.

—Charles Haddon Spurgeon, 1871

The art of being wise is the art of knowing what to overlook.

—Williams James, 1890

Knowledge, a rude unprofitable mass,
The mere materials with which wisdom builds […].
Knowledge is proud that he has learned so much,
Wisdom is humble that he knows no more.

—William Cowper, 1785

Knowledge comes, but wisdom lingers.

—Alfred Tennyson, 1835[3]

---

[1]This maxim, misattributed to quite a few recent authors, is often extended in the following direction: "…and full morons do not learn even by their own errors". (On a personal note, with no hope of reaching the first category, I was always clinging to the second of them, making every effort to avoid the third one.).

[2]There is a close Italian proverb, "Your enemy makes you wise".

[3]Frequently misattributed to Calvin Coolidge.

K. K. Likharev (ed.), *Essential Quotes for Scientists and Engineers*,
https://doi.org/10.1007/978-3-030-63332-5_7

Knowledge for knowledge's sake, anyway - is the worst of all. [...] Knowledge should lead to wisdom, and [...] if it doesn't, it's just a disgusting waste of time!

—J. D. Salinger, 1957

The saddest aspect of life right now is that science gathers knowledge faster than society gathers wisdom.

—Isaac Asimov, 1988

[In paleontology,] increasing knowledge leads to triumphant loss of clarity.

—Alfred Romer, 1962

Where expertise prevails, wisdom vanishes.

—Erwin Chargaff, 1980

Wisdom is what's left after we've run out of personal opinions.

—Cullen Hightower[4]

It is vital to remember that information – in the sense of raw data – is not knowledge, that knowledge is not wisdom, and the wisdom is not foresight. But information is the first step to all these.

—Arthur C. Clarke, 2003

***

The folly of mistaking a paradox for a discovery, a metaphor for a proof, a torrent of verbiage for a spring of capital truths, and oneself for an oracle, is inborn in us.

—Paul Valéry, 1895

The trouble about trying to make yourself stupider than you really are is that you very often succeed.

—C. S. Lewis, 1955

Against stupidity the gods themselves contend in vain.

—Friedrich von Schiller, 1801

Never approach a cow from the front, of a horse from the rear, and a fool from any direction.

—European proverb[5]

Earth has its boundaries, but human stupidity is limitless.

—Gustave Flaubert, 1880

Genius may have its limitations, but stupidity is not thus handicapped.

—Elbert Hubbard, 1906

---

[4]As quoted, without a date, in several reliable collections, starting not later than 1988.

[5]It is curious that in different languages, different animals are mentioned—"cow" in the Russian version, "bull" the English one, "goat" in Yiddish, etc. However, fools look very international.

Real stupidity beats artificial intelligence every time.

—Terry Pratchett, 1996

Egotism is the anesthetic that dulls the pain of stupidity.

—Knute Rockne, 1949[6]

It is not easy to show the road to a blind man.

—Italian proverb

It is difficult to get a man to understand something when his job depends on not understanding it.

—Upton Sinclair, 1935

A great deal of intelligence can be invested in ignorance when the need for illusion is deep.

—Saul Bellow, 1976

No matter what side of the argument you are on, you always find people on your side that you wish were on the other.

—Jascha Heifetz[7]

All you need in this life is ignorance and confidence; then success is sure.

—Mark Twain, 1887

The fundamental cause of trouble is that [...] the stupid are cocksure while the intelligent are full of doubt.

—Bertrand Russell, 1935

I am patient with stupidity but not with those who are proud of it.

—Edith Sitwell, 1967

Nothing is more terrible than to see ignorance in action.

—Johann Wolfgang von Goethe, 1826

***

Brutes by their natural instinct have produced many discoveries, whereas men by discussion and the conclusions of reason have given birth to few or none.

—Francis Bacon, 1620

People who have read a good deal rarely make great discoveries. I do not say this to excuse laziness, but because invention presupposes an extensive contemplation of things on one's own account.

—Georg Christoph Lichtenberg[8]

---

[6]This aphorism is frequently misattributed to Frank Leahy, in whose 1955 book it appeared (with a reference to K. Rockne).

[7]As quoted, without a date, in several reliable collections, starting not later than 1987.

[8]From his *Notebook E* (1775-1776).

There is nobody so irritating as somebody with less intelligence and more sense than we have.

—*Don Herold*, 1924

Beware of the man who works hard to learn something, learns it, and finds himself no wiser than before. [...] He is full of murderous resentment of people who are ignorant without having come by their ignorance the hard way.

—*Kurt Vonnegut*, 1963

***

Those who wish to appear wise among fools, among the wise seem foolish.

—*Quintilian*, ∼95 AD

Were there no fools, there would be no wise men.

—*German proverb*

Let us be thankful for the fools. But for them the rest of us could not succeed.

—*Mark Twain*, 1897

Be wiser than other people, if you can; but do not tell them so.

—*Lord Chesterfield*, 1745

***

Ability will never catch up with the demand for it.

—*Malcolm Forbes*, 1990

There is something that is much more scarce, something rarer than ability. It is the ability to recognize ability.

—*Robert Half*, 1960

It requires wisdom to understand wisdom: the music is nothing if the audience is deaf.

—*Walter Lippmann*, 1929

The most basic question is not what is best, but who shall decide what is best.

—*Thomas Sowell*, 1980

***

There is no great genius without some touch of madness.

—*Seneca*[9]

We are obliged to regard many of our original minds as crazy — at least until we have become as clever as they are.

—*Georg Christian Lichtenberg*[10]

---

[9]This is a succinct paraphrase (with a reference) of an earlier line by Aristotle.
[10]From his *Notebook D* (1773-1775).

Sometimes a scream is better than a thesis.

—*Ralph Waldo Emerson*, 1836

A little Madness in the Spring
Is wholesome even for the King.[11]

—*Emily Dickinson*, 1875

Sanity is a cozy lie.

—*Susan Sontag*, 1961

Perhaps if you know that you are insane then you are not insane.

—*Philip K. Dick*, 1962

The only normal people are the ones you don't know very well.

—*Joe Ancis*[12]

Insanity: a perfectly rational adjustment to an insane world.

—*R. D. Laing*, 1972

In a mad world only the mad are sane.

—*Akira Kurosawa*, 1985

A paranoid is someone who knows a little of what's going on.

—*William S. Burroughs*, 1970

I've always found paranoia to be a perfectly defensible position.

—*Pat Conroy*, 2010

I became insane, with long intervals of horrible sanity.

—*Edgar Allan Poe*, 1848

Some people never go crazy. What truly horrible lives they must lead.

—*Charles Bukowski*, 1987

---

[11]This is apparently a twist of an English proverb: "A little nonsense now and then/Is relished by the wisest men".

[12]As quoted, without a date, in several generally reliable collections, starting not later than 1990, i.e. during the author's lifetime.

On Faith and Religion

We walk by faith, not by sight.

—*The Bible* (2 Corinthians 5:7)

Do not seek to understand in order to believe.

—*Saint Augustine*[1]

Reason is the greatest enemy that faith has.

—*Martin Luther*, 1569

***

Faith may be defined briefly as an illogical belief in the occurrence of the improbable.

—*H. L. Mencken*, 1922

A casual stroll through the lunatic asylum shows that faith does not prove anything.

—*Anonymous*[2]

If the only way you can accept an assertion is by faith, then you are conceding that it can't be taken on its own merits. It is intellectual bankruptcy.

—*Dan Barker*, 1992

Say what you will about the sweet miracle of unquestioning faith, I consider a capacity for it terrifying and absolutely vile.

—*Kurt Vonnegut*, 1966

Belief is the antithesis to thinking.

—*Bergen Evans*, 1946

Belief in the supernatural reflects a failure of the imagination.

—*Edward Abbey*, 1990

---

[1]From his *Tractates on the Gospel of John*, written in the fifth century AD.

[2]This is a popular paraphrase of a much longer line by Friedrich Nietzsche (circa 1888).

K. K. Likharev (ed.), *Essential Quotes for Scientists and Engineers*,
https://doi.org/10.1007/978-3-030-63332-5_8

Pray, *v.* To ask that the laws of the universe be annulled in behalf of a singe petitioner confessedly unworthy.

—Ambrose Bierce, 1911

He who will not reason is a bigot; he who cannot is a fool; and he who dares not is a slave.

—William Drummond, 1805[3]

The fact that a believer is happier than a skeptic is no more to the point than the fact that a drunken man is happier than a sober one.

—George Bernard Shaw, 1913

***

To argue with a person who has renounced the use and authority of reason […] is like administering medicine to the dead.

—Thomas Paine, 1778

Rational arguments don't usually work on religious people. Otherwise, there would be no religious people.

—Doris Egan, 2007

The mind of a bigot is like the pupil of the eye. The more light you shine on it, the more it will contract.

—Oliver Wendell Holmes Sr., 1831

***

If you talk to God, you are praying. If God talks to you, you have schizophrenia.

—Thomas Szasz, 1973

We could call order by the name of God, but it would be an impersonal God. There's not much personal about the laws of physics.

—Stephen Hawking, 2002

I have too much respect for the idea of God to make it responsible for such an absurd world.

—Georges Duhamel, 1937

I don't know if God exists, but it would be better for His reputation if He didn't.

—Jules Renard, 1925

The only excuse for God is that he does not exist!

—Stendhal[4]

It is the final proof of God's omnipotence that he need not exist in order to save us.

—Peter De Vries, 1958

---

[3]Frequently misattributed to his namesake (but "of Hawthornden" rather than "of Logiealmond"), also a Scottish poet, who lived much earlier (1585-1649).

[4]As quoted, without a date, in many sources - e.g., in the well-known biography collection by James Huneker in 1908.

The impotence of God is infinite.[5]

—Anatole France, 1925

***

Not only is there no God, but try getting a plumber on weekends.

—Woody Allen, 1969

It is bad luck to be superstitious.

—Anonymous[6]

Every creed promises a paradise which will be absolutely uninhabitable for anyone of civilized taste.

—Evelyn Waugh, 1942

***

A belief in a supernatural source of evil is not necessary; men alone are quite capable of every wickedness.

—Joseph Conrad, 1911

Don't let us make imaginary evils, when you know we have so many real ones to encounter.

—Oliver Goldsmith, 1768

To such heights of evil has religion been able to drive men.

—Lucretius, first century BC

Men never do evil so completely and cheerfully as when they do it from religious conviction.

—Blaise Pascal, 1670

There is no crime, which men have not committed under the idea of pleasing the Divinity, or appeasing his wrath.

—Baron d'Holbach, 1772

Preparations for the slaughter of mankind have always been made in the name of God.

—Jaroslav Hašek, 1921

We have just enough religion to make us hate, but not enough to make us love one another.

—Jonathan Swift, 1706

---

[5]This joke, as well as the four ones just before it, come close to a major line of serious critical inquiry of religion, whose beginning may be traced back at least to Epicurus: "Is God willing to prevent evil, but not able? Then he is not omnipotent. Is he able, but not willing? Then he is malevolent. Is he both able and willing? Then whence cometh evil? Is he neither able nor willing? Then why call him God?" Unfortunately, this collection is too short to trace this inquiry at earnest.

[6]This quote is frequently attributed to some Andrew W. Mathis, but I could not find a reliable confirmation of this authorship, or even identify this person.

Anyone who can make you believe absurdities, can make you commit atrocities.

—Voltaire, 1765

Religion is an insult to human dignity. With or without it, you would have good people doing good things and evil people doing evil things. But for good people to do evil things, that takes religion.

—Steven Weinberg, 1999

***

Science, by its definition, must exclude the idea of a personal and active providence.

—Henry Adams, 1919

There can be no truce between science and religion.

—J. B. S. Haldane, 1923

If we are honest - and scientists have to be - we much admit that religion is a jumble of false assertions, with no basis in reality.

—Paul Dirac, 1927

The effort to reconcile science and religion is almost always made not by theologians, but by scientists unable to shake off altogether the piety absorbed with their mother's milk.

—H. L. Mencken, 1956

***

Scriptures, *n.* The sacred books of our holy religion, as distinguished from the false and profane writings on which all other faiths are based.

—Ambrose Bierce, 1911

We are all atheists about most of the gods that societies have ever believed in. Some of us just go one god further.

—Richard Dawkins, 2002

Thanks to God, I'm still an atheist.

—Luis Buñuel, 1959

She believed in nothing; only her skepticism kept her from being an atheist.

—Jean-Paul Sartre, 1964

I do not pretend to know where many ignorant men are sure - this is all that agnosticism means.

—Clarence Darrow, 1925

***

With most men, unbelief in one thing springs from blind belief in another.

—Georg Christoph Lichtenberg[7]

---

[7]From his *Notebook L* (1793–1796).

Even if you don't believe in God, you still have to believe in something that gives meaning to your life, and shapes your sense of the world.

—Michael Crichton, 2004

Man is a credulous animal, and must believe *something*; in the absence of good grounds for belief, he will be satisfied with bad ones.

—Bertrand Russell, 1950

Disbelief in magic can force a poor soul into believing in government and business.

—Tom Robbins, 1976

***

Faith is, at one and the same time, absolutely necessary and altogether impossible.

—Stanisław Lem, 1976

Piety is sweet to infant minds.

—William Wordsworth, 1814

Faith is the consolation of the wretched and the terror of the happy.

—Luc de Clapiers, 1747

God is the immemorial refuge of the incompetent, the helpless, the miserable.

—H. L. Mencken, 1956

It is for the good of the state that man should be deluded by religion.

—Marcus Terentius Varro, ~40 BC

Promises, covenants, and oaths, which are the bonds of human society, can have no hold upon an atheist.

—John Locke, 1689

It is [...] absolutely necessary for princes and people to have deeply engraved in their minds the notion of a Supreme Being, creator, ruler, remunerator, and avenger. [...] If God did not exist, it would be necessary to invent him.

—Voltaire, 1770

Our Constitution was made only for a moral and religious people. It is wholly inadequate for the government of any other.

—John Adams, 1798

How many daily read the Word,
And still from vice are not deterred.

—European proverb

I do not believe that, on the balance, religious belief has been a force for good.

—Bertrand Russell, 1928

The greatest injury of all was done by basing morals on myth. For, sooner or later, myth is recognized for what it is, and disappears. Then morality loses the foundation on which it has been built.

—Herbert Samuel, 1947

It is time that science, having destroyed the religious basis for morality, accepted the obligation to provide a new and rational basis for human behavior - a code of ethics concerned with man's needs on earth, not his rewards in heaven.

—Barney Oliver, 1972

Fed on the dry husks of facts, the human heart has a hidden want which science cannot supply.

—William Osler, 1904

Who - aside from certain big children who are indeed found in the natural sciences - still believes that the findings of astronomy, biology, physics, or chemistry could teach us anything about the *meaning* of the world?

—Max Weber, 1922

We have to deal with [...] moral phenomena as products of evolution.

—Herbert Spencer, 1897

It doesn't trivialize morality to argue that it is based on evolution and secular reason.

—Jerry Coyne, 2015[8]

---

[8]I am sorry I could not trace this novel line of scientific/ethical inquiry at earnest in this small collection.

On Human Attitudes

A positive attitude may not solve all your problems, but it will annoy enough people to make it worth the effort.

—Herm Albright[1]

To err is human, to forgive divine.

—Alexander Pope, 1711[2]

To err is human; to forgive, infrequent.

—Franklin P. Adams, 1946

Injuries may be forgiven, but not forgotten.

—Aesop[3]

I never forgive, but I always forget.

—Arthur Balfour[4]

Do not dig a hole for another one to fall into, you may end up in it yourself.

—Russian proverb

Revenge [is] the sweetest morsel to the mouth, that ever was cooked in hell!

—Walter Scott, 1818

Living well is the best revenge.

—George Herbert, 1640[5]

---

[1]As quoted, without a date, in several generally reliable collections, starting not later than 2005.

[2]This is an extension of one of Plutarch's venerable "morals", which was later (in 1386) twisted along a different line by Geoffrey Chaucer - and certainly many others.

[3]Included, without a date, in all major publications of Aesop's *Fables*.

[4]As quoted, without a date, by R. Blake in 1970.

[5]This is a succinct version of an earlier (circa 1580) aphorism by John Lyly.

© The Author(s), under exclusive license to Springer Nature Switzerland AG 2021
K. K. Likharev (ed.), *Essential Quotes for Scientists and Engineers*,
https://doi.org/10.1007/978-3-030-63332-5_9

There is no revenge so complete as forgiveness.

—Josh Billings, 1873

***

Tact is the knack of making a point without making an enemy.

—Howard W. Newton[6]

In the end, we will remember not the words of our enemies, but the silence of our friends.

—Martin Luther King Jr., 1957

Never explain—your friends do not need it, and your enemies will not believe you anyway.

—Elbert Hubbard, 1907

Thank Lord, my dear, that we have enemies,
This means that probably we have friends as well.

—Yuri Vizbor, 1973[7]

Defend me from my friends; I can defend myself from my enemies.

—Jean Hérault Gourville, 1724

To find a friend one must close one eye. To keep him—two.

—Norman Douglas, 1941

***

I know that there are people who do not love their fellow man, and I hate people like that!

—Tom Lehrer, 1965

There are some people who can receive a truth by no other way than to have their understanding shocked and insulted.

—Carl Sandburg, 1904

Never offend people with style when you can offend them with substance.

—Sam Brown, 1997

Resentment […] is like drinking poison and waiting for the other person to die.

—Carrie Fisher, 2004

Holding anger is a poison. It eats you from the inside.

—Mitch Albom, 2003

---

[6]As quoted, without a date, in several reliable collections, starting not later than 1946; sometimes misattributed to Sir Isaac Newton.

[7]This is a prosaic translation of a verse from a popular Russian song..

Never go to bed mad. Stay up and fight!

—*Phyllis Diller*, 1966[8]

In anger we should refrain both from speech and action.

—*Pythagoras*[9]

Where we have strong emotions, we're liable to fool ourselves.

—*Carl Sagan*, 1980

Intolerance is evidence of impotence.

—*Aleister Crowley*, 1929

Fervor is the weapon of choice for the impotent.

—*Frantz Fanon*, 1952

Rudeness is the weak man's imitation of strength.

—*Eric Hoffer*, 1955

Truth often suffers more by the heat of its defenders, than from the arguments of its opposers.

—*William Penn*, 1682

***

A man must be orthodox upon most things, or he will never even have time to preach his own heresy.

—*G. K. Chesterton*, 1909

It takes too much energy to be against something unless it's really important.

—*Madeleine L'Engle*, 2004

What's the sense of wrestling with a pig? You both get all over muddy ... and the pig likes it.

—*Cyrus S. Ching*, 1948[10]

If you are arguing with an idiot, he is probably doing the same.

—*Mikhail Zhvanetsky*, 2010

Never answer a critic, unless he's right.

—*Bernard Baruch*, 1983

***

---

[8]Sometimes misattributed to William Congreve.

[9]As quoted, without a date, by Diogenes Laërtius in the third century AD.

[10]As quoted in a newspaper of that time; frequently misattributed to George Bernard Shaw.

We judge ourselves by what we feel capable of doing, while others judge us by what we have already done.

—Henry Wadsworth Longfellow, 1849

Whenever I dwell for any length of time on my own shortcomings, they gradually begin to seem mild, harmless, rather engaging little things, not at all like the staring defects in other people's characters.

—Margaret Halsey, 1938

The people who are late are often so much jollier than the people who have to wait for them.

—E. V. Lucas, 1926

There are two types of people—those who come into a room and say, 'Well, here I am!' and those who come in and say, 'Ah, there you are.'

—Frederick Lewis Collins[11]

What the world needs is more geniuses with humility; there are so few of us left.

—Oscar Levant[12]

Humility is the solid foundation of all the virtues.

—Confucius, fifth century BC[13]

Humility is a virtue all preach, none practice, and yet everybody is content to hear.

—John Selden, 1786

People with humility don't think less of themselves, they just think of themselves less.

—Ken Blanchard and Norman Vincent Peale, 1996

Humility is a virtue when you have no other.

—Edward Abbey, 1990

Many a man is praised for his reserve and so-called shyness when he is simply too proud to risk making a fool of himself.

—J. B. Priestley, 1956

***

The man with a plan is king. [...] A man without a plan is just a pawn to circumstance.

—Wheeler L. Baker, 1993

---

[11]Some sources attribute this observation to engineer Archie Frederick Collins, who lived at about the same time, but for me this looks very unlikely.

[12]As quoted, without a date, by Gene Perret in 1995.

[13]Possibly based on a line from the Hebrew *Books of Kings*, whose first written version was already available at Confucius' time.

If you don't have a plan for yourself, you'll be a part of someone else's.

—Anonymous[14]

Adventure is just bad planning.

—Roald Amundsen[15]

Plans are useless, but planning is indispensable.

—Dwight Eisenhower, 1962

I rarely plan my research; it plans me.

—Max Perutz, 1998

Man proposes and God disposes.

—European proverb

The absence of alternatives clears the mind marvelously.

—Henry Kissinger, 1978

***

Courage is the art of being the only one who knows you're scared to death.

—Harold Wilson, 1977

Courage is doing what you're afraid to do. There can be no courage unless you're scared.

—Eddie Rickenbacker, 1977

Many would be cowards if they had courage enough.

—Thomas Fuller, 1732

***

Each of us bears his own Hell.

—Virgil[16]

Most powerful is he who has himself in his own power.

—Seneca, ∼64 AD

To be strong, conquer yourself.

—German proverb

Nothing can bring you peace but yourself.

—Ralph Waldo Emerson, 1841

---

[14] I have seen this quote attributed to Emanuel James "Jim" Rohn (1930-2009), and also used, slightly paraphrased, in a 1996 book by some John Mason, but could not find a reliable confirmation of either authorship.

[15] As quoted, without a date, in several reputable sources, including *Wikiquote*.

[16] From his *Aeneid*, written in 29–19 BC.

Nothing can contribute more to peace of soul than the lack of any opinion whatsoever.

—Georg Christoph Lichtenberg[17]

***

We immediately become more effective when we decide to change ourselves rather than asking things to change for us.

—Stephen Covey, 1989

Don't go around saying the world owes you a living. The world owes you nothing. It was here first.

—Robert Jones Burdette, 1905[18]

What you're supposed to do when you don't like a thing is change it. If you can't change it, change the way you think about it. Don't complain.

—Maya Angelou, 2013

I can't complain, but sometimes I still do.

—Joe Walsh, 1978

I will hear not those who weep and complain, for their disease is contagious.

—Og Mandino, 1968

Never complain. Complaints will always discredit you.

—Baltasar Gracián, 1647

Never complain and never explain.

—Stanley Baldwin, 1943[19]

***

Nobody got anywhere in the world by simply being content.

—Louis L'Amour, 1979

Discontent is the wheel that moves people forward.

—Lu Xun[20]

Restlessness is discontent —and discontent is the first necessity of progress. Show me a thoroughly satisfied man and I will show you a failure.

—Thomas Edison, 1948

To be pleased with one's limits is a wretched state.

—Johann Wolfgang von Goethe, 1833

---

[17]From his *Notebook E* (1775-1776).

[18]Frequently misattributed to Mark Twain.

[19]This maxim is sometimes attributed to Benjamin Disraeli (S. Baldwin's predecessor as a UK prime minister), but I could not find a reliable confirmation of that authorship.

[20]As quoted, without a date, in many collections.

Hungry fighters win fights.

—Norman Mailer, 1960

Normal is the wrong name often used for the average.

—Henry S. Haskins, 1940

Normal is getting dressed in clothes that you buy for work and driving through traffic in a car that you are still paying for—in order to get to the job you need to pay for the clothes and the car, and the house you leave vacant all day so you can afford to live in it.

—Ellen Goodman[21]

The trouble with normal is it always gets worse.

—Bruce Cockburn, 1983

Normal is not something to aspire to, it's something to get away from.

—Jodie Foster, 2004

Men who never get carried away should be.

—Malcolm Forbes, 1979

Anxiety is the dizziness of freedom.

—Søren Kierkegaard, 1844

If everything seems under control, you're not going fast enough.

—Mario Andretti[22]

---

[21] As quoted, without a date, in several generally reliable collections, including *Wikiquote.*

[22] He is a famous race car driver.

On Optimism, Pessimism, and Happiness

An optimist stays up to see the New Year in. A pessimist waits to make sure the old one leaves.

—Bill Vaughan[1]

The optimist proclaims that we live in the best of all possible worlds; and the pessimist fears this is true.

—James Branch Cabell, 1926[2]

Idealism is what precedes experience; cynicism is what follows.

—David T. Wolf[3]

***

It is difficult not to write satire.

—Juvenal[4]

That power of accurate observation is commonly called cynicism by those who have not got it.

—George Bernard Shaw, 1894

Cynic, *n.* A blackguard whose faulty vision sees things as they are, not as they ought to be.

—Ambrose Bierce, 1906

Cynicism is an unpleasant way of saying the truth.

—Lillian Hellman, 1939

Being cynical is the only way to deal with modern civilization.

—Frank Zappa, 1982

---

[1] As quoted, without a date, in several reliable collections, including *Wikiquote*.

[2] Sometimes misattributed to J. Robert Oppenheimer.

[3] As quoted, without a date, in several generally reliable collections, starting not later than 1987.

[4] From his *Satires*, written in the early second century AD.

K. K. Likharev (ed.), *Essential Quotes for Scientists and Engineers*,
https://doi.org/10.1007/978-3-030-63332-5_10

No matter how cynical you get, it's never enough to keep up.

—Jane Wagner, 1985[5]

The nice part about being a pessimist is that you are constantly being either proven right or pleasantly surprised.

—George Will, 1994

If you keep saying things are going to be bad, you have a good chance of being a prophet.

—Isaac Bashevis Singer[6]

Everything that can go wrong, will.[7]

—Anonymous

If anything just can't go wrong, it will anyway.

—Francis P. Chisholm, 1963

Walls sink and dunghills rise.

—European proverb

***

Who could live without hope?

—French proverb

Two worst foes of man's existence [are] Fear and Hope.

—Johann Wolfgang von Goethe, 1832

Hope is a good breakfast, but it is a bad supper.

—Francis Bacon, 1624

Hope deceives more men than cunning does.

—Luc de Clapiers, 1746

Hope, like faith, is nothing if it is not courageous; it is nothing if it is not ridiculous.

—Thornton Wilder, 1967

A man without hope is a man without fear.

—Frank Miller, 1986

---

[5]This line (as well as all other quotes by Jane Wagner in this collection) is frequently attributed to Lily Tomlin - the actress it was written for.

[6]As quoted, without a date, in several generally reliable sources, starting not later than 2003.

[7]This is my favorite formulation of the so-called *Murphy Law*, reportedly based on an oral, more personal statement by some Edward A. Murphy made in the 1940s, and first quoted by several authors in the early 1950s. Of its innumerous later forms and twists, I especially like the *1st Chisholm Law* quoted next, its corollary, "Anytime things appear to be going better, you have overlooked something", and the ultimately succinct (anonymous?) maxim: "Constants aren't".

I hope for nothing.
I fear nothing.
I am free.

—Nikos Kazantzakis, 1957[8]

Blessed is he who expects nothing, for he shall never be disappointed.

—Alexander Pope, 1727[9]

Only on the firm foundation of unyielding despair, can the soul's habitation henceforth be safely built.

—Bertrand Russell, 1910

***

I find nothing more depressing than optimism.

—Paul Fussell, 1990

Optimism [...] is the obstinacy of maintaining that everything is best when it is worst.

—Voltaire, 1759

Optimist, *n.* A proponent of the doctrine that black is white.

—Ambrose Bierce, 1911

The place where optimism most flourishes is the lunatic asylum.

—Havelock Ellis, 1923

Positive, *adj.* Mistaken at the top of one's voice.

—Ambrose Bierce, 1911

Only fools are positive.

—Anonymous[10]

When they come downstairs from their Ivory Towers, idealists are very apt to walk straight into the gutter.

—Logan Pearsall Smith, 1931

At least two thirds of our miseries spring from human stupidity, human malice, and those great motivators and justifiers of malice and stupidity: idealism, dogmatism and prosely-tizing zeal on behalf of religious or political idols.

—Aldous Huxley, 1952

***

---

[8]His self-epitaph.

[9]A. Pope declared this was a ninth Beatitude, added by him to the eight ones listed in the Bible (Matthew 5:3–12).

[10]I have seen this quote attributed to actor Moe Howard, but could not find a reliable confirmation of his authorship.

One does not, sir, prove oneself a superior man by perceiving the world in an odious light.

—François-René de Chateaubriand, 1802

It is easier to blame than do better.

—German proverb

How much easier it is to be critical than to be correct.

—Benjamin Disraeli, 1860

Any fool can criticize, condemn, and complain—and most fools do.

—Dale Carnegie, 1936

***

Cynics are right nine times out of ten; what undoes them is their belief that they are right ten times out of ten.

—Charles Issawi, 1973[11]

An idealist believes the short run doesn't count. A cynic believes the long run does not matter. A realist believes that what is done (or left undone) in the short run determines the long run.

—Sidney J. Harris, 1979

I'm a pessimist about probabilities, I'm an optimist about possibilities.

—Lewis Mumford[12]

The essence of my optimism is constructive pessimism.

—Fausto Cercignani, 2004

No man who is correctly informed as to the past, will be disposed to take a morose or desponding view of the present.

—Thomas Babington Macaulay, 1861

I have always been—I think any student of history almost inevitably is—a cheerful pessimist.

—Jacques Barzun[13]

There are two ways to slide easily through life; to believe everything or to doubt everything. Both ways save us from thinking.

—Alfred Korzybski, 1921

***

Happiness is an imaginary condition, formerly attributed by the living to the dead, now usually attributed by adults to children, and by children to adults.

—Thomas Szasz, 1973

---

[11]The first clause of this statement is frequently misattributed to H. L. Mencken.

[12]As quoted, without a date, by Carey Winfrey in 1977.

[13]As quoted, without a date, by Thomas Vinciguerra in 2006.

Happiness isn't something you experience; it's something you remember.

—*Oscar Levant*, 1972

The true paradises are the paradises that we have lost.

—*Marcel Proust*, 1927[14]

Happiness is nothing more than good health and a bad memory.

—*Albert Schweitzer*[15]

Happiness, *n*. An agreeable sensation arising from contemplating the misery of others.[16]

—*Ambrose Bierce*, 1906

Our happiness or unhappiness depends more upon the way in which we meet the events of life, than upon the nature of those events themselves.

—*Wilhelm von Humboldt*, 1849[17]

A large income is the best recipe for happiness I ever heard of.

—*Jane Austen*, 1814

Money doesn't always bring happiness. People with ten million dollars are no happier than people with nine million dollars.[18]

—*Anonymous*[19]

If happiness truly consisted in physical ease and freedom from care, then the happiest individual would not be a man or a woman, it would be, I think, an American cow.

—*William Lyon Phelps*, 1927

It is better to be a human being dissatisfied than a pig satisfied.

—*John Stuart Mill*, 1861

Happiness in intelligent people is the rarest thing I know.

—*Ernest Hemingway*, 1986

A person is never happy except at the price of some ignorance.

—*Anatole France*, 1923[20]

To be stupid, selfish, and have good health are three requirements for happiness, though if stupidity is lacking, all is lost.

—*Gustave Flaubert*, 1846

---

[14]Frequently paraphrased as "The only paradise is the paradise lost".

[15]As quoted in several generally reliable sources, starting not later than 1960.

[16]Looks like a development of a much earlier line by François de La Rochefoucauld: "We all have strength enough to endure the misfortune of others".

[17]Sometimes misattributed to his younger brother, naturalist Alexander von Humboldt.

[18]Apparently a reaction to the common proverb "Money can't buy happiness".

[19]Frequently attributed to Hobart Brown, but I was unable to find a reliable confirmation of his authorship.

[20]The date of the apparently first English publication.

The greatest happiness you can have is knowing that you do not necessarily require happiness.

—*William Saroyan*, 1939

The pursuit of happiness is a most ridiculous phrase; if you pursue happiness you'll never find it.

—*Carrie Snow*[21]

Happiness is not achieved by the conscious pursuit of happiness; it is generally the by-product of other activities.

—*Aldous Huxley*, 1945

The secret of being miserable is to have leisure to bother about whether you are happy or not. The cure for it is occupation.

—*George Bernard Shaw*, 1910

To abandon the struggle for private happiness, to expel all eagerness for temporary desire, to burn with passion for eternal things—this is emancipation.

—*Bernard Russell*, 1903

I cannot believe that the purpose of life is to be happy. I think the purpose of life is to be useful, [...] to have made some difference that you lived at all.

—*Leo Calvin Rosten*, 1965

The only true happiness comes from squandering ourselves for a purpose.

—*John Mason Brown*, 1947

Nobody of [the scientists] knew exactly what is happiness and what constitutes the meaning of life. So they have accepted a working hypothesis that happiness is in a continuing exploration of the unknown, and the meaning of life is in the same.

—*Arkady and Boris Strugatsky*, 1965

---

[21]As quoted, without a date, in several reputable collections; sometimes misattributed to C. P. Snow.

On Speech and Silence, Truth and Lies

When all other means of communication fail, try words.

—*Ashleigh Brilliant*, mid-1970s

***

He who knows does not speak; he who speaks does not know.

—*Lǎozi*, ~ fourth century BC

Wise men talk because they have something to say; fools, because they have to say something.

—*Plato*, first century AD

Saying what we think gives us a wider conversational range than saying what we know.

—*Cullen Hightower*[1]

Just imagine the silence in the world, if people talked only what they knew.

—*Karel Čapek*, 1936

***

Word is a shadow of a deed.

—*Democritus*[2]

For if your meaning's threatened with stagnation,
Then words come in, to save the situation.[3]

—*Johann Wolfgang von Goethe*, 1808

---

[1]As quoted, without a date, in many sources, starting not later than 1998.

[2]As quoted, without a date, in several reliable sources, including Kathleen Freeman's collection published in 1948.

[3]Translation by Philip Wayne. This verse (from Part 1 of *Faust*) is frequently paraphrased shorter: "When ideas fail, words come in handy".

K. K. Likharev (ed.), *Essential Quotes for Scientists and Engineers*,
https://doi.org/10.1007/978-3-030-63332-5_11

Don't *say* things. What you *are* stands over you a while, and thunders so that I cannot hear what you say to the contrary.

—*Ralph Waldo Emerson*, 1876

Speech is conveniently located midway between thought and action, where it often substitutes for both.

—*John Holmes*, 1927

I like business because it rewards deeds and not words.

—*William Feather*, 1927

It is a common delusion that you make things better by talking about them.

—*Rose Macaulay*, 1926

I know of only one bird, the parrot, that talks; and it can't fly very high.

—*Wilbur Wright*, 1908[4]

***

Long talk makes short days.

—*French proverb*

But far more numerous was the herd of such,
Who think too little and who talk too much.

—*John Dryden*, 1681

The secret of being a bore is to tell everything.

—*Voltaire*, 1738

A healthy male adult bore consumes each year one and a half times his own weight in other people's patience.

—*John Updike*, 1965

Talk low, talk slow, and don't talk too much.

—*John Wayne*[5]

Make sure you have finished speaking before your audience has finished listening.

—*Dorothy Sarnoff*, 1970

The secret of a good sermon is to have a good beginning and a good ending, then having the two as close together as possible.

—*George Burns*[6]

---

[4]The aviation pioneer has reportedly made this statement at a banquet, declining an invitation to give a talk. This was apparently a paraphrase of an earlier maxim by Sakya Pandita, but I like it much more than the original.

[5]As quoted, without a date, in many sources, including J. Bartlett's collection.

[6]Possibly based on a more succinct German proverb (sometimes attributed to Martin Luther), "The fewer the words, the better the prayer".

For the wise, one word is enough.

—*European proverb*[7]

***

For the wise, silence is an answer.

—*Euripides*[8]

Silence is a friend who will never betray.

—*Confucius*, fifth century BC

Speech is silver, silence is gold.

—*European proverb*

Drawing on my fine command of the English language, I said nothing.

—*Robert Benchley*[9]

One of the lessons of history is that nothing is often a good thing to do, and always a clever thing to say.

—*Will Durant*, 1958

Blessed is the man who, having nothing to say, abstains from giving us wordy evidence of the fact.

—*George Eliot*, 1879

The fool who is silent passes for the wise.

—*European proverb*

Better to remain silent and be thought a fool than to speak out and remove all doubt.

—*Anonymous*[10]

Look wise, say nothing, and grunt. Speech was given to conceal thought.

—*William Osler*[11]

Well-timed silence has more eloquence than speech.

—*Martin Farquhar Tupper*, 1849

Silence is the most perfect expression of scorn.

—*George Bernard Shaw*, 1921

Silence is the virtue of a fool.

—*Francis Bacon*, 1605

---

[7]May be traced back to the Latin "*Verbum Sat Sapienti*".

[8]As quoted, without a date, in several reputable collections, including *Wikiquotes*.

[9]As quoted, without a date, by C. E. Sylvester in 2005.

[10]Frequently misattributed to either Samuel Johnson, or Abraham Lincoln, or Mark Twain.

[11]As quoted, without a date, by William Bennett Bean in 1950. The second sentence paraphrases an earlier (circa 1763) line by Voltaire.

Speech is civilization itself. The word, even the most contradictory word, preserves contact - it is silence which isolates.

—*Thomas Mann*, 1924

***

Human beings [...] are very much at a mercy of the particular language which has become the medium of expression for their society.

—*Edward Sapir*, 1929

Language is the source of misunderstandings.

—*Antoine de Saint-Exupéry*, 1943

The great thing about human language is that it prevents us from sticking to the matter at hand.

—*Lewis Thomas*, 1974

Never express yourself more clearly than you are able to think.

—*Niels Bohr*, 1922

When in doubt, mumble.[12]

—*James H. Boren*, 1970

The difference between the almost right word and the right word is really a large matter – 'tis the difference between the lightning bug and the lightning.

—*Mark Twain*, 1888

Broadly speaking, the short words are the best, and the old words, when short, best of all.

—*Winston Churchill*, 1949

When there is a gap between one's real and one's declared aims, one turns as it were instinctively to long words and exhausted idioms, like a cuttlefish spurting out ink.

—*George Orwell*, 1950[13]

Grasp the subject, the words will follow.

—*Cato,* second century BC

It is feeling and force of imagination that makes us eloquent.

—*Quintilian*, ~95 AD

We believe a scientist because he can substantiate his remarks, not because he is eloquent and forcible in his enunciation. In fact, we distrust him when he seems to be influencing us by his manner

—*I. A. Richards*, 1926

***

---

[12]This is the title of J. Boren's nice little book—a good companion to *Parkinson's Law* and *The Peter Principle* in their description of bureaucracy.

[13]This is, actually, just a more succinct paraphrase of a much earlier (circa 1691) line by John Ray.

A spoken word is not a sparrow. Once it flies out, you can't catch it.

—European proverb

My method [...] is to take the utmost trouble to find the right thing to say, and then to say it with the utmost levity.

—George Bernard Shaw, 1896

No word is ill spoken that is not ill taken.

—European proverb

He who says what he likes, must hear what he does not like.

—European proverb

Always be ready to speak your mind and a base man will avoid you.[14]

—William Blake, 1793

***

The truth shall make you free.

—The Bible (John 8:32)

The truth that makes men free is for the most part the truth which men prefer not to hear.

—Herbert Agar, 1945

The truth? The truth, Lazarus, is perhaps something so unbearable, so terrible, something so deadly, that simple people could not live with it!

—Miguel de Unamuno, 1913

As a rule people are afraid of truth. Each truth we discover in nature or social life destroys the crutches on which we need to lean.

—Ernst Toller, 1944

There are few nudities so objectionable as the naked truth.

—Agnes Repplier, 1891

The first reaction to truth is hatred.

—Tertullian, ∼97 AD

As scarce as truth is, the supply has always been in excess of the demand.

—Josh Billings, 1865

In a time of deceit, telling the truth is a revolutionary act.

—Anonymous[15]

Speak the truth, but leave immediately after.

—Slovenian proverb

---

[14]I can testify that this old recipe still works fine.

[15]Frequently misattributed to George Orwell.

There are only two ways of telling the complete truth - anonymously and posthumously.

—*Thomas Sowell*, 2013

The truth does not change according to our ability to stomach it emotionally.

—*Flannery O'Connor*, 1979

Two is not equal to three, not even for the largest value of two.

—*Anonymous*[16]

He who has no good memory should never take upon him the trade of lying.

—*Michel de Montaigne*, 1595

Deceivers are the most dangerous members of society. They trifle with the best affections of our nature, and violate the most sacred obligations.

—*George Crabbe*, 1816

O, what a tangled web we weave,
When first we practice to deceive!

—*Walter Scott*, 1808

By a lie, a man throws away and, if it were, annihilates his dignity as a man.

—*Immanuel Kant*, 1790

He who does not bellow the truth when he knows the truth, makes himself the accomplice of liars and forgers.

—*Charles Péguy*, 1899

Show me a liar, and I'll show you a thief.

—*European proverb*

To be believable, a lie should be patched with truth.

—*Danish proverb*

A lie which is half a truth is ever the blackest of lies.

—*Alfred Tennyson*, 1864

The only vice that cannot be forgiven is hypocrisy.

—*William Hazlitt*, 1823

Hypocrisy is a revolting, psychopathic state.

—*Anton Chekhov*, 1888

Hypocrisy is the vice of vices [...] Only the hypocrite is really rotten to the core.

—*Hannah Arendt*, 1963

***

---

[16]The so-called *Grabel's Law*, based on a somewhat different joke by Arvin Grabel.

Part of the inhumanity of the computer is that, once it is competently programmed and working smoothly, it is completely honest.

—Isaac Asimov, 1983

There are truths that are not for all men, not for all times.

—Voltaire, 1764

It's a basic truth of the human condition that everybody lies. The only variable is about what.

—David Shore, 2004

Everybody lies, but it doesn't matter because nobody listens.

—Anonymous[17]

One of the best ways to keep a great secret is to shout it.

—Edwin H. Land, 1960

Nobody really listens to anyone else, and if you try it for a while you'll see why.

—Mignon McLaughlin, 1981

Most conversations are simply monologues delivered in the presence of witnesses.

—Margaret Millar, 2012

The opposite of talking isn't listening. The opposite of talking is waiting.[18]

—Fran Lebowitz, 1981

---

[17]I have seen this remark attributed to some Nick Diamos, but could not find a reliable confirmation of this authorship.

[18]This is essentially a paraphrase of a much earlier saying by Francois de La Rochefoucauld, but I like it more than the original.

On Writing, Literature, Poetry, and Journalism

Unprovided with original learning, unformed in the habits of thinking, unskilled in the arts of composition, I resolved - to write a book.

—*Edward Gibbon*, 1796

After being Turned Down by numerous Publishers, he had decided to write for Posterity.

—*George Ade*, 1900

Nature seems to have provided that the follies of men should be transient, but they by writing books render them permanent.

—*Montesquieu*, 1721

The only thing I was fit for was to be a writer, and this notion rested solely on my suspicion that I would never be fit for real work, and that writing didn't require any.

—*Russell Baker*, 1983

It took me fifteen years to discover that I had no talent for writing, but I couldn't give it up because by that time I was too famous.

—*Robert Benchley*, 1955

Reading stories is bad enough but writing them is worse.

—*L. M. Montgomery*, 1908

I can't understand why a person will take a year to write a novel when he can easily buy one for a few dollars.

—*Fred Allen*[1]

When I want to read a book, I write one.

—*Benjamin Disraeli*, 1868[2]

All of us learn to write in the second grade. Most of us go on to greater things.

—*Bobby Knight*, 1986

---

[1]As quoted, without a date, in several reputable sources, starting not later than in 1996.

[2]This famous maxim is actually a twist of a much earlier (circa 1824) line by Washington Irving.

© The Author(s), under exclusive license to Springer Nature Switzerland AG 2021   101
K. K. Likharev (ed.), *Essential Quotes for Scientists and Engineers*,
https://doi.org/10.1007/978-3-030-63332-5_12

How vain it is to sit down to write when you have not stood up to live!

—*Henry David Thoreau*, 1851

The only reason for being a professional writer is that you can't help it.

—*Leo Calvin Rosten*, 1972

***

Writing is not necessarily something to be ashamed of, but do it in private and wash your hands afterwards.

—*Robert* A. *Heinlein*, 1973

Writers should be read, but neither seen nor heard.

—*Daphne du Maurier*, 1938

When your work speaks for itself, don't interrupt.

—*Henry J. Kaiser*[3]

Words, once they are printed, have a life of their own.

—*Carol Burnett*, 1981

What is written without effort is in general read without pleasure.

—*Samuel Johnson*, 1799

No tears in the writer, no tears in the reader.

—*Robert Frost*, 1939

Word by word the great books are made.

—*French proverb*

How long does it take to write a good book? All of the years that you've lived.

—*Edward Abbey*, 1990

***

In writing, fidelity to fact leads eventually to the poetry of truth.

—*Edward Abbey*, 1990

The skill of writing is to create a context in which other people can think.

—*Edwin Schlossberg*[4]

Great literature is simply language charged with meaning to the utmost possible degree.

—*Ezra Pound*, 1931

The very essence of literature is the war between emotion and intellect.

—*Isaac Bashevis Singer*, 1985

---

[3]As quoted, without a date, in many sources, starting not later than 1968.

[4]As quoted, without a date, in several generally reliable sources, starting not later than 2006.

In literature as in love, we are astonished by what is chosen by others.

—André Maurois, 1940

In literature as in love, the most interesting parts are between the words.

—Mikhail Zhvanetsky, 2010

Literature is the orchestration of platitudes.

—Thornton Wilder, 1953

What is responsible for the success of many works is the rapport between the mediocrity of the authors' ideas and the mediocrity of the public.

—Nicolas Chamfort, 1796

There's many a best-seller that could have been prevented by a good teacher.

—Flannery O'Connor, 1969

Today's literature: prescriptions written by patients.

—Karl Kraus, 1976

***

This was a book to kill time for those who liked it better dead.

—Rose Macaulay, 1921

The covers of this book are too far apart.

—Ambrose Bierce[5]

In every fat book there is a thin book trying to get out.

—Anonymous[6]

Think much, say little, write less.

—European proverb

Vigorous writing is concise.

—William Strunk Jr., 1918[7]

[Writing,] I try to leave out the parts that people skip.

—Elmore Leonard, 2009

I made this [letter] longer only because I have not had the leisure to make it shorter.

—Blaise Pascal, 1656

---

[5]As quoted, without a date, in several major collections of his work.

[6]Probably based on a similar joke about men by George Orwell (circa 1939).

[7]This is a quote from Strunk's wonderful little book *The Elements of Style*, later extended and revised by his former student E. B. White, and currently available in its fourth edition (1999). See also Dorothy Parker's advice below.

I write slowly [because] writing briefly takes far more time.

—Carl Friedrich Gauss[8]

Anyone who can think clearly can write clearly. But neither is simple.

—William Feather, 1949

If you have any young friends who aspire to become writers, the second-greatest favor you can do them is to present them with copies of *The Elements of Style*. The first-greatest, of course, is to shoot them now, while they're happy.

—Dorothy Parker, 1959

***

Nothing is said nowadays that has not been said before.

—Terence, 161 BC

Nothing from nothing ever yet was born.

—Lucretius, first century BC

There's nothing clever that hasn't been thought of before — you've just got to try to think it all over again.

—Johann Wolfgang von Goethe, 1833

About the most originality that any writer can hope to achieve honestly is to steal with good judgment.

—Josh Billings, 1871

Originality is the fine art of remembering what you hear but forgetting where you heard it.

—Laurence J. Peter, 1977[9]

Immature poets imitate; mature poets steal.

—T. S. Eliot, 1920[10]

Fine words! I wonder where you stole 'em.

—Jonathan Swift, 1724

The substance, the bulk, the actual and valuable material of all human utterances – is plagiarism.

—Mark Twain, 1903

When you steal from one author, it's plagiarism. If you steal from many, it's research.

—Wilson Mizner, 1953

***

---

[8]As quoted, without a date, by George F. Simmons in 1922.

[9]I have seen a similar aphorism attributed to Gene Fowler, but could not find a reliable confirmation of that authorship.

[10]This maxim was later repeatedly reproduced—at least by Lionel Trilling about artists (in 1962) and by Igor Stravinsky about composers (in 1967).

History, *n*. An account, mostly false, of events mostly unimportant, which are brought about by rulers mostly knaves, and soldiers mostly fools.

—Ambrose Bierce, 1906

What then is [...] the truth of history? A fable agreed upon.

—Napoléon, 1816

The history [is] written by the victors.

—Alexis Guignard, 1842[11]

It will be much better by all Parties to leave the past to history, especially as I propose to write the history myself.[12]

—Winston Churchill, 1948

Ignorance is the first requisite of the historian - ignorance, which simplifies and clarifies, which selects and omits, with a placid perfection unattainable by the highest art.

—Lytton Strachey, 1918

Though God cannot alter the past, historians can.

—Samuel Butler, 1901

***

I am constantly writing autobiography, but I have to turn it into fiction in order to give it credibility.

—Katherine Paterson, 1989

Of all forms of fiction, autobiography is most gratuitous.

—Tom Stoppard, 1966

Autobiography - that unrivaled vehicle for telling the truth about other people.

—Philip Guedalla, 1923

***

No one sees his own faults.

—Spanish proverb

Isolation in a creative work is an onerous thing. It is better to have negative criticism than none at all.

—Anton Chekhov, 1886

Asking a working writer what he thinks about critics is like asking a lamppost what it feels about dogs.

—Christopher Hampton, 1977

---

[11]Widely misattributed to various later authors—most frequently to Napoléon, and to Winston Churchill who merely loved to quote it.

[12]This quip is frequently paraphrased in a shorter form: "History will be kind to me for I intend to write it.".

Criticism is prejudice made plausible.

—*H. L. Mencken*, 1919

No degree of dullness can safeguard a work against the determination of critics to find it fascinating.

—*Harold Rosenberg*, 1959

The remarkable thing about Shakespeare is that he really is very good - in spite of all the people who say he is very good.

—*Robert Graves*, 1964

Literature is strewn with the wreckage of men who have minded beyond reason the opinions of others.

—*Virginia Woolf*, 1929

I cannot give you the formula for success, but I can give you the formula for failure: try to please everyone.

—*Herbert Bayard Swope*, 1950

Write something to suit yourself and many people will like it; write something to suit everybody and scarcely anyone will care for it.

—*Jesse Stuart*, 1949

Better to write for yourself and have no public, than to write for the public and have no self.

—*Cyril Connolly*, 1933

***

I would define, in brief, the Poetry of words as *The Rhythmical Creation of Beauty.*

—*Edgar Allan Poe*, 1850

Poetry is to prose as dancing to walking.

—*John Wain*, 1976

Writing free poetry is like playing tennis with the net down.

—*Robert Frost*, 1935

One possible definition of our modern culture is that it is one in which nine-tenths of our intellectuals can't read any poetry.

—*Randall Jarrell*, 1953

Publishing a volume of verse is like dropping a rose petal down the Grand Canyon and waiting for an echo.

—*Don Marquis*, 1962

We cannot foresee how would our word echo,
And empathy comes to us as the Lord's grace.[13]

—Fyodor Tyutchev, 1869

Most people ignore most poetry
because
most poetry ignores most people.

—Adrian Mitchell, 1964

Poets have been mysteriously silent on the subject of cheese.

—G. K. Chesterton, 1909

Poetry is indispensable – if I only knew what for.

—Jean Cocteau, 1959

In science one tries to tell people, in such a way as to be understood by everyone, something that no one ever knew before. But in poetry, it's the exact opposite.

—Paul Dirac, 1958

Good! He did not have enough imagination for mathematics.

—David Hilbert[14]

I have nothing to say, and I am saying it, and that is poetry.

—John Cage, 1949

Poetry often enters through the window of irrelevance.

—M. C. Richards, 1962

The office of Poetry is not to make us think accurately, but feel truly.

—Frederick William Robertson, 1852

***

The current opinion that science and poetry are opposed is a delusion. [...] Those who have never entered upon scientific pursuits know not a tithe of the poetry by which they are surrounded.

—Herbert Spencer, 1855

The opposition between science and feeling is largely a misunderstanding.

—Arthur Thomson, 1921

Natural knowledge has not forgone emotion. It has simply taken for itself new ground of emotion, under impulsion from and in sacrifice to that one of its 'values', Truth.

—Charles Scott Sherrington, 1940

---

[13]The Russian original is a wonderful (not "free"!) verse, for which I could not find (or produce :-) a fair poetic translation to English.

[14]On the news that one of his students has dropped out to study poetry; as quoted by David J. Darling in 2004.

Far more marvelous is the truth than any artists of the past imagined! Why do the poets of the present not speak of it?

—Richard Feynman, 1964

Whoever knows many things by nature is a poet.

—Pindar, 476 BC

A man should be learned in several sciences, and should have a reasonable, philosophical and in some measure a mathematical head, to be a complete and excellent poet.

—John Dryden, 1674

Thus easier 'tis to hold that many things
Have primal bodies in common (as we see
The single letters common to many words)
[...]
Vexed through the ages (as indeed they are)
By the innumerable blows of chance.

—Lucretius, first century BC[15]

In his hand
He took the golden compasses prepar'd
In God's eternal store, to circumscribe
This Universe, and all created things:
One foot he centr'd, and the other turned
Round through the vast profunditie obscure,
And said, 'Thus far extend, thus farr they bounds,
This be thy just Circumference, O World.'

—John Milton, 1667[16]

To see a World in a Grain of Sand,
And a Heaven in a Wild Flower,
Hold Infinity in the palm of your hand,
And Eternity in an hour.

—William Blake, 1863[17]

I am no poet, but if you think for yourselves, as I proceed, the facts will form a poem in your minds.

—Michael Faraday, 1858

Mendeleev's Periodic Table [is] poetry, loftier and more solemn than all the poetry we had swallowed down in liceo; and come to think of it, it even rhymed!

—Primo Levi, 1975

---

[15]From his famous poem *On the Nature of Things*; translation by William Ellery Leonard.
[16]From *Paradise Lost*.
[17]From *Auguries of Innocence*.

Science arose from poetry, and [...] a change of times might beneficially reunite the two as friends, at a higher level and to mutual advantage.

—Johann Wolfgang von Goethe, 1817

***

Literature is news that STAYS news.

—Ezra Pound, 1934

People everywhere confuse what they read in newspapers with news.

—A. J. Liebling, 1956

For most folks, no news is good news; for the press, good news is not news.

—Gloria Borger, 1995

The window to the world can be covered by a newspaper.

—Stanisław Jerzy Lec, 1957

The public have an insatiable curiosity to know everything, except what is worth knowing. Journalism, conscious of this, and having tradesman-like habits, supplies their demands.

—Oscar Wilde, 1891

A newspaper [...] consists of just the same number of words, whether there be any news in it or not.

—Henry Fielding, 1749

Charles: 'Anything interesting in *The Times*?'
Ruth: 'Don't be silly, Charles.'

—Noël Coward, 1941

The Times is getting more detestable (but that is too weak word) than ever.

—Charles Darwin, 1863

Whenever I read *Time* or *Newsweek* or such magazines, I wash my hands afterward. But how to wash off the small but odious stain such reading leaves on the mind?

—Edward Abbey, 1990

The man who never looks into a newspaper is better informed than those he who reads them. [...] He who reads nothing will still learn the great facts, and the details are all false.

—Thomas Jefferson, 1807

Once a newspaper touches a story, the facts are lost forever, even to the protagonists.

—Norman Mailer, 1960

To read a newspaper is to refrain from reading something worthwhile. The first discipline of education must therefore be to refuse resolutely to feed the mind with canned chatter.

—Aleister Crowley, 1929

The biases the media has are much bigger than conservative or liberal. They're about getting ratings, about taking money, about doing stories that are easy to cover.

—Al Franken, 2002

Rock journalism is people who can't write, interviewing people who can't talk, for people who can't read.[18]

—Frank Zappa, 1993

The making of a journalist: no ideas and the ability to express them.

—Karl Kraus, 1976

Every journalist has a novel in him, which is an excellent place for it.

—Russell Lynes, 1966

Editor: a person employed by a newspaper, whose business it is to separate the wheat from the chaff, and to see that the chaff is printed.

—Elbert Hubbard, 1913[19]

Some editors are failed writers - but so are most writers.

—T. S. Eliot, 1982

---

[18]I believe that the actual applicability of this definition is much broader.

[19]A similar, but evidently later line by Adlai E. Stevenson is frequently quoted.

On Arts, Entertainment, and Internet

The perfection of art is to conceal art.

—Quintilian, ∼95 AD

The highest condition of art is artlessness.

—Henry David Thoreau, 1854

I myself do nothing. The Holy Spirit accomplishes all through me.

—William Blake[1]

Art is a collaboration between God and the artist, and the less the artist does the better.

—André Gide, 1950

The position of the artist is humble. He is essentially a channel.

—Piet Mondrian, 1942

I saw the angel in the marble and carved until I set him free.

—Anonymous[2]

It is the artist's business to create sunshine when the sun fails.

—Romain Rolland, 1908

***

A great artist is always before his time or behind it.

—George Moore, 1929

Imitation is suicide.

—Ralph Waldo Emerson, 1841

---

[1] As quoted, without a date, in several generally reliable sources, starting not later than 1997.
[2] Broadly attributed to Michelangelo. Indeed his Sonnet 15 (circa 1550) carries a similar idea, but in a longer form.

© The Author(s), under exclusive license to Springer Nature Switzerland AG 2021
K. K. Likharev (ed.), *Essential Quotes for Scientists and Engineers*,
https://doi.org/10.1007/978-3-030-63332-5_13

In art, there are only two types of people: revolutionaries or plagiarists.

—*Paul Gauguin*, 1895

Art is vice. You don't marry it legitimately, you rape it.

—*Edgar Degas*, 1918

Art is a communication of ecstasy.

—*Minor White*, 1950[3]

Art is a form of catharsis.

—*Dorothy Parker*, 1944

Art is a revolt against fate.

—*André Malraux*, 1951

Art is not truth. Art is a lie that makes us realize truth, at least the truth that is given to us to understand.

—*Pablo Picasso*, 1972

Art has nothing to do with taste.

—*Max Ernst*[4]

Art is anything you can get away with.

—*Marshall McLuhan*, 1964

Art is making something out of nothing and selling it.

—*Frank Zappa*, 2003

Any fool can paint a picture, but it takes a wise man to sell it.

—*Samuel Butler*, 1950[5]

[Abstract art is] a product of the untalented, sold by the unprincipled to the utterly bewildered.

—*Al Capp*, 1963

How many people become abstract in order to appear profound!

—*Joseph Jourbert*[6]

One reassuring thing about modern art is that things can't be as bad as they are painted.

—*Anonymous*[7]

***

---

[3]Sometimes misattributed to P. D. Ouspensky.

[4]As quoted, without a date, by Jürgen Pech in 1996.

[5]I am not confident that this quip had not been published earlier.

[6]As quoted, without a date, by M. Paul de Raynal in 1862.

[7]I have seen this observation attributed to some M. Walthall Jackson, but was unable to confirm this authorship.

There is only one admirable form of the imagination: the imagination that is so intense that it creates a new reality, that it makes things happen, whether it be a political thing or a social thing or a work of art.

—*Sean Ó Faolain*[8]

Imagination is the beginning of creation. You imagine what you desire; you will what you imagine; and at last you create what you will.

—*George Bernard Shaw*, 1921

In the war against Reality man has one weapon - Imagination.

—*Jules de Gaultier*[9]

There is nothing more dreadful than imagination without taste.

—*Johann Wolfgang von Goethe*, 1819

Ah, good taste! What a dreadful thing! Taste is the enemy of creativeness.

—*Pablo Picasso*, 1957

Skill without imagination is craftsmanship and gives us many useful objects such as wickerwork picnic baskets. Imagination without skill gives us modern art.

—*Tom Stoppard*, 1972

Only a fool is scornful of the commonplace.

—*W. Somerset Maugham*, 1939

Without the aid of prejudice and custom, I should not be able to find my way across the room.

—*William Hazlitt*, 1852

No one is so keen to gather ever new impressions as those who do not know how to process the old ones.

—*Marie von Ebner-Eschenbach*, 1893

In painting as in eloquence, the greater your strength, the quieter will be your manner.

—*John Ruskin*, 1860

***

Aesthetic value is often the by-product of the artist striving to do something else.

—*Evelyn Waugh*, 1976

There are two kinds of art: (1) decorative, nonobjective, wallpaper art; and (2) art with a moral purpose.

—*Edward Abbey*, 1990

---

[8]As quoted, without a date, in several reputable sources, starting not later than 1978, i.e. during author's lifetime.

[9]As quoted, without a date, in many sources, starting not later than 1935, i.e. during author's lifetime.

All art is propaganda.

—*Upton Sinclair*, 1925

***

Architecture, of all the arts, is the one which acts the most slowly, but the most surely, on the soul.

—*Ernest Dimnet*, 1932

An arch never sleeps.

—*James Fergusson*, 1910[10]

Architecture is frozen music.

—*Friedrich Schelling*, 1803[11]

Gothic cathedrals and Doric temples are mathematics in stone.

—*Oswald Spengler*, 1923

The physician can bury his mistakes, but the architect can only advise his client to plant vines.

—*Frank Lloyd Wright*, 1953

***

After silence, that which comes nearest to expressing the inexpressible is music.

—*Aldous Huxley*, 1931

Music expresses that which cannot be put into words and that which cannot remain silent.

—*Victor Hugo*, 1864

Without music, life would be a mistake.[12]

—*Friedrich Nietzsche*, 1888

If there is a Kingdom of Heaven, it lies in music.

—*Edward Abbey*, 1989

The function of music is to release us from the tyranny of conscious thought.

—*Thomas Beecham*[13]

---

[10]The author acknowledged a similar Hindu proverb.

[11]A similar aphorism is frequently attributed to Goethe, but the first publication of his version dates somewhat later (1829).

[12]This line evidently refers to the music known in author's times, nowadays called the "classical music".

[13]As quoted in his biography by H. Atkins and A. Newman (1978).

'Tis good; though music oft hath such a charm
To make bad good, and good provoke to harm.

—William Shakespeare, 1604

Extraordinary how potent cheap music is.

—Noël Coward, 1930

I occasionally play works by contemporary composers and for two reasons. First to discourage the composer from writing any more and secondly to remind myself how much I appreciate Beethoven.

—Jascha Heifetz, 1961

'Rock': music to hummer out fenders by.

—Edward Abbey, 1990

The worst wheel makes the most noise.

—European proverb

What looked like it might have been some kind of counterculture is, in reality, just the plain old chaos of undifferentiated weirdness.

—Jerry Garcia[14]

***

Let us go singing as far as we go: the road will be less tedious.

—Virgil, 37 BC

He who sings drives away the sorrow.

—Italian proverb

It is the best of all trades to make songs, and the second best is to sing them.

—Hilaire Belloc, 1909

Nowadays what isn't worth saying is sung.

—Pierre Beaumarchais, 1775

Opera is when a guy gets stabbed in the back and, instead of bleeding, he sings.

—Ed Gardner[15]

I don't mind what language an opera is sung in so long as it is a language I don't understand.

—Edward Appleton, 1955

---

[14]Note that he was not some old curmudgeon, but one of rock music's pillars.

[15]From his *Duffy's Tavern* radio program series (1941-1951).

Writing about music is like dancing about architecture.

—*Martin Mull*, 1979[16]

***

The little foolery that wise men have makes a great show.

—*William Shakespeare*, 1600

Good taste is the enemy of comedy.

—*Mel Brooks*[17]

Good taste is the worst vice ever invented.

—*Edith Sitwell*, 1967

Drama is life with dull bits cut out.

—*Alfred Hitchcock*, 1960

I write plays because writing dialogue is the only respectable way of contradicting yourself. [...] I put a position, rebut it, refute the rebuttal, and rebut the refutation.

—*Tom Stoppard*, 1977

If you want to see your plays performed the way you wrote them, become President.

—*Václav Havel*, 1990[18]

Theater supposes lives that are poor and agitated, a people searching in dreams a refuge from thought.

—*Romain Rolland*, 1903

If the play is good, there is no need to bother actors: reading it is sufficient to get the proper impression. And if the play is bad, no acting can make it good.

—*Anton Chekhov*, 1889

Actors are a necessary evil.

—*Alfred Hitchcock*, 1943

If you want to help the American Theater, don't be an actress, darling. Be an audience.

—*Tallulah Bankhead*, 1952

Don't put your daughter on the stage, Mrs Worthington,
Don't put your daughter on the stage.

—*Noël Coward*, 1935

We're actors – we're the opposite of people.

—*Tom Stoppard*, 1967

---

[16]Frequently misattributed to Elvis Costello.

[17]As quoted, without a date, in several generally reliable collections, starting not later than 1991.

[18]Before becoming the President of Czechoslovakia in 1989 (and then the first President of the Czech Republic in 1993), he was a prominent playwright.

Acting is merely the art of keeping a large group of people from coughing.

—Ralph Richardson, 1946

Acting is the most minor of gifts and not a very high-class way to earn a living. Shirley Temple could do it at the age of four.

—Katharine Hepburn[19]

Acting is not being emotional, but being able to express emotion.[20]

—Kate Reid, 1988

Acting is a form of deception, and actors can mesmerize themselves as easily as an audience.

—Leo Calvin Rosten, 1941

Honesty. That's the main thing in the theater today. Honesty... and just as soon as I can learn how to fake it, I'll have it made.

—Celeste Holm, 1962[21]

Even supposing a young man of appreciable mental powers to be lured upon the stage, [...] his mind would be inevitably and almost immediately destroyed by the gaudy nonsense issuing from his mouth every night.

—H. L. Mencken, 1920

***

You can take all the sincerity in Hollywood, place it in the navel of a firefly and still have room enough for three caraway seeds and a producer's heart.

—Fred Allen, 1959

The modern film [...] requires no contribution from the audience but a mouthful of popcorn.

—Raymond Chandler, 1962

While our people is still illiterate, the most important arts for us are cinema and circus.

—Vladimir Lenin, 1921

There is only one thing that can kill the Movies, and that is education.

—Will Rogers, 1949

***

---

[19] As quoted, without a date, in several generally reliable collections, starting not later than 1993.

[20] "To express" or "to fake"?—see also the next two quotes.

[21] A similar line is attributed to Daniel Schorr, but its first record is only circa 1992. Later, these quips were twisted by many, including George Burns.

The radio and television [...] have succeeded in lifting the manufacture of banality out of the sphere of handicraft and placed it in that of a major industry.

—Nathalie Sarraute, 1960

All television is children's television.

—Anonymous[22]

Television is not the truth! Television is the goddamned amusement park!

—Paddy Chayefsky, 1976

Television: a medium. So called because it is neither rare nor well done.

—Ernie Kovacs, 1996

Television is a rat race, and remember this, even if you win you are still a rat.

—Jackie Gleason, 1956[23]

Television is for appearing on – not for looking at.[24]

—Noël Coward, 1956

I find television very educational. Every time somebody switches it on, I go into another room and read a good book.

—Groucho Marx[25]

Television is the first truly democratic culture – the first culture available to everybody and entirely governed by what the people want. The most terrifying thing is what people do want.

—Clive Barnes, 1969[26]

***

We've heard that a million monkeys at a million keyboards could produce the complete works of Shakespeare; now, thanks to the Internet, we know that is not true.

—Robert Wilensky, 1996

Internet is so big, so powerful and pointless that for some people it is a complete substitute for life.

—Andrew Brown[27]

Information on the Internet is subject to the same rules and regulations as conversation at a bar.

—George D. Lundberg, 1997

---

[22]I have seen this wisdom attributed to Richard Adler, but could not find a reliable confirmation of his authorship.

[23]Later paraphrased/generalized by many, most famously by Lily Tomlin.

[24]This reminds me of prestigious scientific journals like *Science* and *Nature* that are, similarly, for being published in—not for reading of.

[25]As quoted, without a date, in many sources, possibly starting from Leslie Halliwell in 1984.

[26]I believe that this evaluation is even more applicable to the Internet—read on.

[27]As quoted, without a date, in several generally reliable sources, starting not later than 1996.

Doing research on the Web is like using a library assembled piecemeal by pack rats and vandalized nightly.

—Roger Ebert, 1998

Caution: Do not mistake the Internet for an encyclopedia, and the search engine for a table of contents. The Internet is a sprawling databank that's about one-quarter wheat and three-quarters chaff.

—The Associated Press (Stylebook and Briefing on Media Law), 2007[28]

Anyone can post messages to the net. Practically everyone does. The resulting cacophony drowns out serious discussion.

—Clifford Stoll, 1995

Most of us employ the Internet not to seek the best information, but rather to select information that confirms our prejudices.

—Nicholas Kristof, 2009

If the Internet has given us anything, it's some idea how much psychosis goes undiagnosed.

—Jan Burke, 2012

The Internet is like a big circus tent full of scary, boring creatures and pornography.

—Richard Kyanka, 2005

My favorite thing about the Internet is that you get to go into the private world of real creeps without having to smell them.

—Penn Jillette, 2007

It was now impossible to distinguish a roomful of people working diligently from a roomful of people taking the *What-Kind-of-Dog-Am-I?*online personality quiz.

—Rainbow Rowell, 2011

On the Internet, nobody knows you're a dog.

—Peter Steiner, 1993

---

[28]Note the date. As time goes on, proportions change fast.

On Love, Family, and Children

One word
Frees us of all the weight and pain of life:
That word is love.

—Sophocles, ∼406 BC

To love is to be delighted by the happiness of another.

—Gottfried Leibniz, 1677

Fathers and teachers, I ponder, 'What is hell?' I maintain that this is the suffering of being unable to love.

—Fyodor Dostoevsky, 1880

If you have not loved,
You have neither lived nor breathed.

—Vladimir Vysotsky, 1975[1]

Love conquers all.

—Virgil, 37 BC

Love rules without law.

—Italian proverb

***

Oh, what a heaven is love! Oh, what a hell!

—Thomas Dekker, 1604

Reason and love are sworn enemies.

—Pierre Corneille, 1631

---

[1]This is a fair but prosaic translation of a verse from a famous Russian song.

© The Author(s), under exclusive license to Springer Nature Switzerland AG 2021
K. K. Likharev (ed.), *Essential Quotes for Scientists and Engineers*,
https://doi.org/10.1007/978-3-030-63332-5_14

If Jack's in love, he's no judge of Jill's beauty.

—Benjamin Franklin, 1748

Fire in the heart sends smoke into the head.

—German proverb

When the heart speaks, the mind finds it indecent to object.

—Milan Kundera, 1984

The brain and the heart act upon each other in the manner of an hour-glass. One fills itself only to empty the other.

—Jules Renard, 1925

Love is the triumph of imagination over intelligence.

—H. L. Mencken, 1920

Love is only a dirty trick played on us to achieve continuation of the species.

—W. Somerset Maugham, 1938

Love is an exploding cigar we willingly smoke.

—Lynda Barry, 1983

Love is a snowmobile racing across the tundra and then suddenly it flips over, pinning you underneath. At night, the ice weasels come.

—Matt Groening, 1990

Love is a child of illusion and the parent of disillusion.

—Miguel de Unamuno, 1913

When misfortune knocks the door, love flies out of the window.

—European proverb

If love is the answer, could you please rephrase the question?

—Jane Wagner, 1979

Love has a lot of press-agenting from the oldest times; but there are higher, nobler things than love.

—P. G. Wodehouse, 1922

It is better in some respects to be admired by those with whom you live, than to be loved by them, [...] because admiration is so much more tolerant than love.

—Arthur Helps, 1871

Love is like the measles; we all have to go through it. Also like the measles, we take it only once.

—Jerome K. Jerome, 1886

***

There is no reciprocity. Men love women. Women love children, Children love hamsters.

—Alice Thomas Ellis, 1987[2]

Nothing takes the taste out of peanut butter quite like unrequited love.

—Charles M. Schulz[3]

'Cause when love is gone, there's always justice.
And when justice is gone, there's always force.
And when force is gone, there's always Mom.
Hi Mom!

—Laurie Anderson, 1981

Love is a kind of warfare.

—Ovid, 2 AD

You can be in love with a woman and still hate her.

—Fedor Dostoevsky, 1880

If you begin by sacrificing yourself to those you love, you will end by hating those to whom you have sacrificed yourself.

—George Bernard Shaw, 1903

There is no love without jealousy.

—European proverb

In jealousy, there is more of self-love than of love.

—François de La Rochefoucauld, 1664

Jealous, *adj.* Unduly concerned about the preservation of that which can be lost only if not worth keeping.

—Ambrose Bierce, 1906

***

Love makes time pass away, and time makes love pass away.

—French proverb

The promise of mutual exclusive and everlasting love is a promise that cannot be kept and should not be made.

—Havelock Ellis, 1928

A good relationship is like fireworks: loud, explosive, and liable to maim you if you hold on too long.

—Jeph Jacques, 2008

***

---

[2]Sometimes misattributed to Peter Greenway, in whose movie *8½ Women* (1999) it was partly quoted, without a reference.

[3]As quoted, without a date, in many sources, including the author's 1997 compilation of his *Peanuts* comic strips.

Love, *n*. A temporary insanity curable by marriage.

—Ambrose Bierce, 1906

When a Man should marry? A young Man not yet, an Elder Man not at all.

—Francis Bacon, 1625

It is most unwise for people in love to marry.

—George Bernard Shaw, 1903

One should always be in love. That is the reason one should never marry.

—Oscar Wilde, 1907

I wish someone would have told me that, just because I'm a girl, I don't have to get married.

—Marlo Thomas[4]

Many a man who falls in love with a dimple make the mistake of marrying the whole girl.

—Evan Esar, 1943

You have married a beauty? So much worse for you.

—Italian proverb

When a girl marries, she exchanges the attentions of all the other men of her acquaintance for the inattention of just one.

—Helen Rowland, 1903[5]

Marriage is a bribe to make a housekeeper think she's a householder.

—Thornton Wilder, 1954

Marriage is a great institution. [...] But I ain't ready for an institution yet.

—Mae West, 1933

Marriage is a desperate thing.

—John Selden, 1689

Marriage, a market which has nothing free but the entrance.

—Michel de Montaigne, 1580

Marriage isn't a word, it's a sentence.

—King Vidor, 1928

Wedlock is padlock.

—English proverb

The trouble with wedlock is that there's not enough wed and too much lock.

—Christopher Morley, 1939

---

[4]As quoted, without a date, in several generally reliable collections, starting not later than 2007.

[5]A later paraphrase of this quote by K. Hepburn is also popular.

Home life as we understand it is no more natural to us than a cage to a cockatoo.

—George Bernard Shaw, 1908

Marriage is like life in this – that it is a field of battle, and not a bed of roses.

—Robert Louis Stevenson, 1881

All married couples should learn the art of battle as they should learn the art of making love.

—Ann Landers, 1968

My toughest fight was with my first wife.

—Muhammad Ali, 2009

My parents only had one argument in forty-five years. It lasted forty-three years.

—Cathy Ladman[6]

***

The highest task for a bond between two people: that each protects the solitude of the other.

—Rainer Maria Rilke, 1904

God created man and, finding him not sufficiently alone, gave him a companion to make him feel his solitude more keenly.

—Paul Valéry, 1943

Men have a much better time of [marriage] than women. For one thing, they marry later. For another thing, they die earlier.

—H. L. Mencken, 1949

***

A married man with a family will do anything for money.

—Charles de Talleyrand[7]

A married man is half a man.

—Norman Douglas, 1917

Wife and children, believe me, is a big evil,
Of which everything foul in our lives originates.[8]

—Alexander Pushkin, 1822

God gave women intuition and femininity. Used properly, the combination easily jumbles the brain of any man I've ever met.

—Farrah Fawcett, 2000

---

[6]As quoted, without a date, in many collections, starting not later than 1990.

[7]As quoted, without a date, in several collections, starting not later than 1977.

[8]In Russian this is a rhymed, metered verse, for which I could not find a fair poetic translation to English.

The male is a domestic animal which, if treated with firmness and kindness, can be trained to do most things.

–Jilly Cooper, 2006

The only thing worse than a man you can't control is a man you can.

—Margo Kaufman, 1999

Women want mediocre men, and men are working hard to become as mediocre as possible.

—Margaret Mead, 1958

Being a husband is a whole-time job. This is why so many husbands fail. They cannot give their entire attention to it.[9]

—Arnold Bennett, 1918

Women should be praised, whether truly or falsely.

—German proverb

What a woman wills, God wills.

—European proverb

Be to her virtues very kind.
Be to her faults a little blind.

—Matthew Prior, 1707

Marriage is give and take. You'd better give it to her or she'll take it anyway.

—Joey Adams, 1957

***

The greatest mistake is trying to be more agreeable than you can be.

—Walter Bagehot, 1876

Beware the fury of a patient man.

—John Dryden, 1681

A husband is like a fire, he goes out when unattended.

—Evan Esar, 1943

Eighty percent of married men cheat in America. The rest cheat in Europe.

—Jackie Mason, 2002

A man can sleep around, no questions asked, but if a woman makes nineteen or twenty mistakes she's a tramp.

—Joan Rivers, 2001

There is nothing like a good dose of another woman to make a man appreciate his wife.

—Clare Booth Luce, 1936

---

[9]The reader may like to relate this wisdom to Ivan Pavlov's maxim (circa 1936): "Science demands of a man all his life".

Marriage should be celebrated as the optimistic and glorious thing that it is. We can't call it a failure if it doesn't last forever.

—Jeanette Winterson, 2002

Divorces are made in heaven.[10]

—Oscar Wilde, 1895

It is hard to part only with your first husband; then they streak by as mile signs.

—Mikhail Zhvanetsky, 2010

***

The happiest moments of my life have been the few which I have past at home in the bosom of my family.

—Thomas Jefferson, 1790

Happiness is having a large, loving, caring, close-knit family in another city.

—George Burns, 1990

***

Familiarity breeds contempt – and children.[11]

—Mark Twain, 1935

A baby is God's opinion that the world should go on.

—Carl Sandburg, 1948

Of children as of procreation – the pleasure momentary, the posture ridiculous, the expenses damnable.

—Evelyn Waugh, 1954

We want better reasons for having children than not knowing how to prevent them.

—Dora Russell, 1925

To bring into the world an unwanted human being is as antisocial an act as murder.

—Gore Vidal, 1968

***

Somewhere on this globe, every ten seconds, there is a woman giving birth to a child. She must be found and stopped.

—Sam Levenson[12]

---

[10]This is a twist of the European proverb "Marriages are made in heaven", which may be traced back to at least a (longer) line, circa 1567, by William Painter.

[11]The first clause is due to Aesop (fifth century BC).

[12]Included, without a date, into a 2016 compilation of Levenson's aphorisms.

The entire planet and virtually every nation is already vastly overpopulated.

—Paul R. Ehrlich, 1990

Any overpopulation which [...] erodes the world's material resources or its resources of beauty and intellectual satisfaction are evil.

—Julian Huxley, 1975

Every increase in population means a decrease in [dignity, decency, and opportunity].

—Marya Mannes, 1964

The neglect of [the population control], which in existing states is so common, is a never-failing cause of poverty among the citizens.

—Aristotle, fourth century BC

The hungry world cannot be fed until and unless the growth of its resources and the growth of its population come into balance.

—Lyndon B. Johnson, 1966

Beyond a critical point within a finite space, freedom diminishes as numbers increase. [...] The human question is not how many can possibly survive within the system, but what kind of existence is possible for those who do survive.

—Frank Herbert, 1965

***

There are too many people, and too few human beings.

—Robert Zend[13]

We need to make a world in which fewer children are born, and in which we take better care of them.

—George Wald[14]

Most people have no business having children. They are unqualified, either genetically or culturally or both, to reproduce such sorry specimens as themselves.

—Edward Abbey, 1990

The right to live does not connote the right of each man to reproduce his kind. [...] As we lessen the stringency of natural selection [...] we must increase the standard, mental and physical, of parentage.

—Karl Pearson, 1912

Parentage is a very important profession, but no test of fitness for it is ever imposed in the interest of the children.

—George Bernard Shaw, 1944

---

[13] As quoted, without a date, in several generally reputable collections.

[14] As quoted, without a date, in several collections, starting not later than 1986.

Nothing you do for children is ever wasted. They seem not to notice us, hovering, averting our eyes, and they seldom offer thanks, but what we do for them is never wasted.

—Garrison Keillor, 1987

***

Little children, big problems; larger children, even larger problems.

—Anonymous[15]

Before I got married I had six theories about bringing up children; now I have six children and no theories.

—John Wilmot[16]

The thing that impresses me the most about America is the way parents obey their children.

—Prince Edward, 1957

I have found the best way to give advice to your children is to find out what they want and then advise them to do it.

—Harry S. Truman, 1955

Setting a good example for children takes all the fun out of middle age.

—William Feather, 1949

Always be nice to your children because they are the ones who will choose your rest home.

—Phyllis Diller, 2005

The first duty towards children is to make them happy.

—Charles Buxton, 1883

A happy childhood has spoiled many a promising life.

—Robertson Davies, 1985

Few misfortunes can befall a boy which bring worse consequences than to have a really affectionate mother.

—W. Somerset Maugham, 1896

Children need love, especially when they do not deserve it.

—Harold S. Hulbert

Who loves well, chastises well.

—European proverb

Ask your child what he wants for dinner only if he's buying.

—Fran Lebowitz, 1981

---

[15]There is a superficially close Danish proverb, but with the opposite punch line.

[16]This is a rather questionable attribution, without a date, in a 1946 publication by *Family Life Bureau.*

The best way to teach your kids about taxes is by eating 30% of their ice cream.

—Anonymous[17]

There is no such thing as a normal child.

—Lauretta Bender, 1954[18]

Television has changed the American child from an irresistible force into an immovable object.

—Laurence J. Peter, 1977

What music is more enchanting than the voices of young people, when you can't hear what they say?

—Logan Pearsall Smith, 1933

Like its politicians and its wars, society has the teenagers it deserves.

—J. B. Priestley, 1982

Few things are more satisfying than seeing your own children have teenagers of their own.

—Doug Larson[19]

By the time a man realizes that maybe his father was right, he usually has a son who thinks he's wrong.

—Charles Wadsworth[20]

The young always have the same problem – how to rebel and conform at the same time. They have now solved it by defying their parents and copying one another.

—Quentin Crisp, 1968

We are born charming, frank and spontaneous and must be civilized before we are fit to participate in society.

—Judith Martin, 1979

***

Boyhood, like measles, is one of those complaints which a man should catch young and have done with, for when it comes in middle life it is apt to be serious.

—P. G. Wodehouse, 1922

Boys will be boys.

—William Thackeray, 1848

---

[17]Broadly misattributed to actor Bill Murray.

[18]Note that she was a prominent child neuropsychiatrist.

[19]As quoted, without a date, in several generally reliable collections, starting not later than 2003, i.e. during the author's lifetime.

[20]He was a Philadelphia preacher, a friend of Emily Dickinson; this quote is sometimes misattributed to his namesake, a present-day classical pianist.

Boys will be boys, an' so will lots o' ole men.

—Kin Hubbard, 1910[21]

You can only be young once. But you can always be immature.

—Dave Barry, 1995

The secret of eternal youth is arrested development.

—Alice Roosevelt Longworth, 1979

Athletic sports, save in the case of young boys, are designed for idiots.

—George Jean Nathan, 1931

***

The four stages of man are infancy, childhood, adolescence and obsolescence.

—Art Linkletter, 1965

Youth is a blunder; Manhood a struggle; Old Age a regret.

—Benjamin Disraeli, 1844

Youth would be an ideal state if it came a little later in life.

—H. H. Asquith, 1923

No wise man ever wished to be younger.

—Jonathan Swift, 1742

Life begins at forty.

—Walter B. Pitkin, 1932

By the age of forty, a man is responsible for his face. And his fate.

—Edward Abbey, 1990

A fool at forty is fool indeed.

—Edward Young, 1728

There is no fool like an old fool.

—English proverb

Old cow believes that she never was a calf.

—French proverb

The denunciation of the young is a necessary part of the hygiene of older people, and greatly assists in the circulation of their blood.

—Logan Pearsall Smith, 1931

The dead might as well try to speak to the living as the old to the young.

—Willa Cather, 1922

---

[21]The original I have found is signed "Wallace", but from the style and source similarity, I believe this was just one of Kin Hubbard's numerous pen names.

On Money and Wealth

He who knows that enough is enough, will always have enough.

—Lăozi, ~ fourth century BC

It is not the man who has little, but the man who craves for more, that is poor.

—Seneca, ~ 64 AD

That man is richest whose pleasures are the cheapest.

—Henry David Thoreau, 1856

That who has less than he wants, should know that he has more than he deserves.

—Georg Christoph Lichtenberg[1]

Who wants everything - and immediately, will get nothing - and gradually.

—Mikhail Zhvanetsky, 2010

Some luck lies in not getting what you wanted but getting what you have, which once you have got it you may be smart enough to see is what you would have wanted had you known.

—Garrison Keillor, 1985

Anyone whose needs are small seems threatening to the rich, because he's always ready to escape their control.

—Nicolas Chamfort, 1812

Who has nothing, fears nothing.

—European proverb

---

[1]From his *Reflections*, whose apparently first English translation (by Norman Alliston) was published only in 1908.

133
K. K. Likharev (ed.), *Essential Quotes for Scientists and Engineers*,
https://doi.org/10.1007/978-3-030-63332-5_15

You cannot take a cow from a man who has none.

—Danish proverb[2]

***

Poverty is not a vice, but it is a great inconvenience.

—European proverb[3]

To be poor and independent is nearly an impossibility.

—William Cobbett, 1829

The trouble with being poor is that it takes up all of your time.

—Willem de Kooning[4]

Those who have some means think that the most important thing in the world is love. The poor know this is money.

—Gerald Brenan, 1978

Money is better than poverty, if only for financial reasons.

—Woody Allen, 1976

I've been poor and I've been rich. Rich is better!

—Beatrice Kaufman, 1937[5]

[The two most beautiful English words] are 'check' and 'enclosed'.

—Dorothy Parker, 1932

The only way not to think about money is to have a great deal of it.

—Edith Wharton, 1905

There are few sorrows, however poignant, in which a good income is of no avail.

—Logan Pearsall Smith, 1931

Who has, is.
Who has not, is not.
Who has something, is something.

—Italian proverbs

Money couldn't buy friends, but you got a better class of enemy.

—Spike Milligan, 1963

***

You want as much again as you have already got.

—Horace, ~21 BC

---

[2]There are proverbs with the same idea in many other languages.

[3]Could originate from a John Florio's line in his *Second Fruits* (circa 1591).

[4]As quoted, without a date, in several collections, already during author's lifetime.

[5]Often misattributed (among many others) to Sophie Tucker.

Who is rich? He that is content.
Who is that? Nobody.

—Benjamin Franklin, 1755

To be content with little is difficult; to be content with much, impossible.

—Marie von Ebner-Eschenbach, 1893

With wishing comes grieving.

—Italian proverb

Who accepts, sells himself.

—Italian proverb

The ways by which you may get money almost without exception lead downward.

—Henry David Thoreau, 1863

All paid jobs absorb and degrade the mind.

—Aristotle, fourth century BC

Money often costs too much.

—Ralph Waldo Emerson, 1860

Money is a good servant, but a bad master.

—French proverb

Money is the fruit of evil as often as the root of it.

—Henry Fielding, 1731

Acquisition of wealth for its own sake is disgusting.

—Robert Bunsen[6]

Salary is no object: I want only enough to keep my body and soul apart.

—Dorothy Parker, 1928

The art of living easily as to money, is to pitch your scale of living one degree below your means.

—Henry Taylor, 1848

***

All progress is based upon a universal innate desire on the part of every organism to live beyond its income.

—Samuel Butler, 1912

Expenditure rises to meet income.

—C. Northcote Parkinson, 1960

Riches cause arrogance; poverty, meekness.

—German proverb

---

[6]As quoted, without a date, by R. Oesper in 1975.

There is nothing more demoralizing than a small but adequate income.

—*Edmund Wilson*, 1946

Excess on occasion is exhilarating. It prevents moderation from acquiring the deadening effect of a habit.

—*W. Somerset Maugham*, 1938

I spent a lot of money on booze, birds and fast cars. The rest I just squandered.

—*George Best*, 2005

***

The best things in life are free. The second best things are very expensive.

—*Anonymous*

What costs nothing is worth nothing.

—*Dutch proverb*

Every time you spend money, you're casting a vote for the kind of world you want.

—*Anna Lappé*, 2003

Ninety percent of everything is crud.[7]

—*Theodore Sturgeon*, 1957

As long as people accept crap, it will be financially profitable to dispense it.

—*Dick Cavett*, 1971

Commerce is the school for cheating.

—*Luc de Clapiers*, 1747

The best plan is to profit by the folly of others.

—*Pliny*, 77 AD[8]

When fools go to market, peddlers make money.

—*European proverb*

Words fine and bold
Are goods half sold.

—*German proverb*

Advertising may be described as the science of arresting the human intelligence long enough to get money from it.

—*Stephen Leacock*, 1924

---

[7]This popular *Sturgeon's Law* is a jocular embodiment of the serious *Pareto Principle* in applied statistics. Another popular name of this principle is the *20/80 Rule*, stemming from another quasi-joke: "20% of all people drink 80% of all beer".

[8]Note the date. Despite its age, this formula seems to describe many (most?) modern corporate *business plans* pretty well.

Advertising is a valuable economic factor because it is the cheapest way of selling goods, particularly if the goods are worthless.

—Sinclair Lewis, 1922

Give them quality. That's the best advertising in the world.

—Milton Hershey, 1953[9]

Good wine needs no crier.

—European proverb

Unethical advertising uses falsehoods to deceive the public; ethical advertising uses truth to deceive the public.[10]

—Vilhjalmur Stefansson, 1964

If a man deceives you once time, he is a rascal; if he does it twice, you are a fool.

—European proverb

***

Finance, *n.* The art or science of managing revenues and resources for the best advantage of the manager.

—Ambrose Bierce, 1906.

Finance is the art of passing money from hand to hand until it finally disappears.

—Anonymous[11]

The world of finance is a mysterious world in which [...] evaporation precedes liquidation.

—Joseph Conrad, 1915

Banking establishments are more dangerous than standing armies.

—Thomas Jefferson, 1816

***

Behind every great fortune there is a great crime.

—Anonymous[12]

When a man tells you that he got rich through hard work, ask him: 'Whose?'.

—Don Marquis[13]

---

[9]Note the date. Perhaps at that time the words "Hershey" and "quality" were compatible.

[10]It seems like this classification is applicable to most newspaper articles and politicians' speeches (and very unfortunately even some scientific papers) as well.

[11]This quote is frequently attributed to Robert W. Sarnoff (in 1965-1969, the head of the *RCA Corporation*), but I could not find a reliable confirmation of this authorship.

[12]Perhaps based on a much longer line by Honoré de Balzac (circa 1834).

[13]As quoted, without a date, by many, probably starting with Edward Anthony in 1962.

You may know how little God thinks of money by observing on what bad and contemptible characters He often bestows it.

—Thomas Guthrie, 1865[14]

Nothing is more admirable than the fortitude with which millionaires tolerate the disadvantages of their wealth.

—Rex Stout, 1937

The only thing I like about rich people is their money.

—Nancy Astor, 1955

Money-getters are the benefactors of our race. To them, in a great measure, are we indebted for our institutions of learning and of art, our academies, colleges and churches.

—P. T. Barnum, 1880

---

[14]Frequently misattributed to Dorothy Parker.

On Human Condition

[…] naught prevents a man
From living a life even worthy of the gods.

—*Lucretius*, first century BC[1]

An honest man is the noblest work of God.

—*Alexander Pope*, 1734

***

Every man is as God has made him, and sometimes a great deal worse.

—*Miguel de Cervantes*, 1615

What a chimera, then, is man! What a novelty, what a monster, what a chaos, what a contradiction, what a prodigy! […] Glory and scum of the universe.

—*Blaise Pascal*, 1670

As I know more of mankind I expect less of them, and am ready now to call a man a *good man* upon easier terms than I was formerly.

—*Samuel Johnson*, 1783

What a ridiculous world ours is, as far as it concerns the mind of men.

—*Michael Faraday*, 1853

The great mistake is that of looking upon men as virtuous, or thinking that they can be made so by laws.

—*Henry Bolingbroke*, 1752

Out of the crooked timber of humanity, no straight thing was ever made.

—*Immanuel Kant*, 1784

---

[1]From *On the Nature of Things*; translation by William Ellery Leonard.

K. K. Likharev (ed.), *Essential Quotes for Scientists and Engineers*,
https://doi.org/10.1007/978-3-030-63332-5_16

I have no faith in human perfectibility.

—Edgar Allan Poe, 1850

Human action can be modified to some extent, but human nature cannot be changed.

—Abraham Lincoln, 1860

You cannot slander human nature; it is worse than words can paint it.

—Charles Haddon Spurgeon, 1889[2]

Men have never been good, they are not good, and they never will be good.

—Karl Barth, 1948

I am never surprised by bad behavior. I expect it.

—Gore Vidal, 2009

***

What makes mankind tragic is not that they are the victims of nature, it is that they are conscious of it.

—Joseph Conrad, 1898

Humankind differs from the animals only by a little, and most people throw that away.

—Confucius, fifth century BC

I am a human, and nothing human is alien to me.[3]

—Terence, ∼ 160 BC

He who makes a beast of himself, gets rid of the pain of being a man.

—Samuel Johnson, 1809

Man is a clever animal who behaves like an imbecile.

—Albert Schweitzer, 1969

Chimpanzees have given me so much. [...] What I have learned from them has shaped my understanding of human behavior.

—Jane Goodall, 2009

The more you learn about the dignity of the gorilla, the more you want to avoid people.

—Dian Fossey, 1985

I think computer viruses should count as life. [...] I think it says something about human nature that the only form of life we have created so far is purely destructive. We've created life in our own image.

—Stephen Hawking, 1994

---

[2]Note that the author was not some anchorite, but rather a very popular Baptist minister, known in his time as *The Prince of Preachers*.

[3]In my experience, this popular aphorism is quoted mostly by the people who want to justify something *bestial* not being alien to them.

If […] man created the devil, he has created him in his own image.

—Fyodor Dostoevsky, 1880

The devil is an optimist if he thinks he can make people meaner.

—Karl Kraus, 1976

The only thing that stops God from sending another flood is that the first one was useless.

—Nicolas Chamfort, 1771

Every man who at forty is not a misanthrope, has never liked mankind.

—Nicolas Chamfort, 1780

The surest sign that intelligent life exists elsewhere in the universe is that none of it has tried to contact us.

—Bill Watterson, 1989[4]

The only thing that scares me more than space aliens is the idea that there aren't any space aliens. We can't be the best that creation has to offer. I pray we're not all there is. If so, we're in big trouble.

—Ellen DeGeneres[5]

***

Science has made us gods even before we are worthy of being men.

—Jean Rostand, 1939

The world has achieved brilliance without conscience. Ours in the world of nuclear giants and ethical infants.

—Omar Bradley, 1948[6]

Our scientific power has outrun our spiritual power. We have guided missiles and misguided men.

—Martin Luther King Jr., 1963

Modern Man is the victim of the very instruments he values most. Every gain in power, every mastery of natural forces, every scientific addition to knowledge, has proved potentially dangerous, because it has not been accompanied by equal gains in self-understanding and self-discipline.

—Lewis Mumford, 1944

Here we stand in the middle of this new world with our primitive brain, attuned to the simple cave life, with terrific forces at our disposal, which we are clever enough to release, but whose consequences we cannot comprehend.

—Albert Szent-Györgyi, 1962

***

---

[4]Sometimes misattributed to Arthur C. Clarke.

[5]Quoted, without a date, in several collections, starting at least from 2000.

[6]Yes Virginia, that very military general!

Would you persuade, speak of Interest, not Reason.

—Benjamin Franklin, 1734

Reason is the newest and the rarest thing in human life, the most delicate child of human history.

—Edward Abbey, 1990

Man is a Reasoning Animal. Such is the claim. [...] In truth, man is incurably foolish.

—Mark Twain, 1939

A provision of endless apparatus, a bustle of infinite enquiry and research, or even the mere mechanical labor of copying, may be employed, to evade and shuffle off real labor - the real labor of thinking.

—Joshua Reynolds, 1784[7]

People do not like to think. If one thinks, one must reach conclusions, and conclusions are not always pleasant.

—Helen Keller, 1913

Most people would die sooner than think - in fact, they do so.

—Bertrand Russell, 1925

Every thought is an exception from the general rule that people do not think.

—Paul Valéry, 1941

Human beings are seventy percent water, and with some the rest is collagen.

—Martin Mull[8]

Fools have been and always will be the majority of mankind.

—Denis Diderot, 1777

The fools are in a terrible overwhelming majority, all the wide world over.

—Henrik Ibsen, 1882

***

Nothing is so firmly believed, as what we least know.

—Michel de Montaigne, 1595

The most costly of all follies is to believe passionately in the palpably not true. It is the chief occupation of mankind.

—H. L. Mencken, 1949

That which has always been accepted by everyone, everywhere, is almost certain to be false.

—Paul Valéry, 1943

---

[7]Thomas A. Edison liked this quip so much that he had its shorter paraphrase posted around his factory.

[8]As quoted, without a date, by several collections, starting not later than 1994.

Most human beings have an almost infinite capacity for taking things for granted.

—Aldous Huxley, 1950

The world wants to be deceived.

—Sebastian Brant, 1494

How easily are people deceived, how do they like prophets and oracles, what a herd they are!

—Anton Chekhov, 1921

You may fool all the people some of the time, and some of the people all of the time, but you can't fool all of the people all of the time.

—Abraham Lincoln[9]

You can fool too many of the people too much of the time.

—James Thurber, 1939

The public will believe anything, so long as it is not founded on truth.

—Edith Sitwell, 1965

Men have always had a greater inclination to close their eyes than to open them, to believe comforting lies rather than discomforting truth.

—Stephen Vizinczey, 1969

Occasionally he stumbled over the truth, but hastily picked himself up and hurried on as if nothing ever happened.[10]

—Winston Churchill, 1936

Three things will never be believed - the true, the probable, and the logical.

—John Steinbeck, 1961

The men the American public admire most extravagantly are the most daring liars; the men they detest most violently are those who try to tell them the truth.

—H. L. Mencken, 1922

I never did *give* anybody hell. I just told the truth, and they *thought* it was hell.[11]

—Harry S. Truman, 1958

If you make people think they're thinking, they'll love you; but if you really make them think, they'll hate you.

—Don Marquis, 1923

***

---

[9]Not found in Lincoln's papers, this famous aphorism was first quoted, without a date, in a 1886 newspaper, to become a commonplace soon after this. (More rarely, it is attributed to P. T. Barnum).

[10]Said about Stanley Baldwin, then the UK Prime Minister.

[11]This was an allusion to his nickname "Give'em Hell Harry".

The great majority of mankind are satisfied with appearances, as though they were realities.

—Niccoló Machiavelli, 1517

Human kind cannot bear much reality.

—T. S. Eliot, 1922

Reality is the leading cause of stress amongst those in touch with it.

—Jane Wagner, 1985

Well, I've wrestled with reality for 35 years, Doctor, and I'm happy to state I finally won out over it.

—Mary Chase, 1950[12]

I believe in looking reality straight in the eye and denying it.

—Garrison Keillor, 1998

Reality is something you rise above.

—Liza Minnelli, 2009

Reality is that which, when you stop believing in it, doesn't go away.

—Philip K. Dick, 1978

---

[12]From her play *Harvey*; pronounced by the main character, Elwood P. Dowd.

On Society and Solitude, Liberty and Civilization

The inventors of genius hasten the march of civilization. The fanatics and hallucinated create history.

—*Gustave Le Bon*, 1895

Indeed history is nothing more than a tableau of crimes and misfortunes.

—*Voltaire*, 1767

I know no study which is so unutterably saddening as that of the evolution of humanity, as it is set forth in the annals of history.

—*Thomas Henry Huxley*, 1889

Happy is the people whose annals in history books are blank.

—*Montesquieu*[1]

History [...] is a struggle between ignorance and injustice.

—*Mikhail Zhvanetsky*, 2010

History teaches us that men and nations behave wisely once they have exhausted all other alternatives.

—*Abba Eban*, 1970

***

In the modern world the intelligence of public opinion is the one indispensable condition for social progress.

—*Charles William Eliot*, 1869

No one in this world, so far as I know [...] has ever lost money by underestimating the intelligence of the great masses of the plain people.

—*H. L. Mencken*, 1926

---

[1]Apparently first quoted, without a date, by Thomas Carlyle in 1864.

© The Author(s), under exclusive license to Springer Nature Switzerland AG 2021
K. K. Likharev (ed.), *Essential Quotes for Scientists and Engineers*,
https://doi.org/10.1007/978-3-030-63332-5_17

Few people are capable of expressing with equanimity opinions which differ from the prejudices of their social environment. Most people are even incapable of forming such opinions.

—Albert Einstein, 1954

There is no nonsense so gross that society will not, at some time, make a doctrine of it and defend it with every weapon of communal stupidity.

—Robertson Davies, 1994

Wherever you come near the human race, there's layers and layers on nonsense.

—Thornton Wilder, 1938

The fact that an opinion has been widely held is no evidence whatever that it is not utterly absurd; indeed in view of the silliness of the majority of mankind, a widespread belief is more likely to be foolish than sensible.

—Bertrand Russell, 1929

The voice of majority is no proof of justice.

—Friedrich von Schiller, 1800

A foolish cry, though taken by thirty-six millions of voices, does not cease to be foolish.

—Anatole France, 1908[2]

Fallacies [...] do not cease to be fallacies because they become fashions.

—G. K. Chesterton, 1930

Public opinion, a vulgar, impertinent, anonymous tyrant who deliberately makes life unpleasant for anyone who is not content to be the average man.

—Dean Inge, 1919

Public opinion is a chaos of superstition, misinformation, and prejudice.

—Gore Vidal, 1965

The pretence of collective wisdom is the most palatable of all impostures.

—William Godwin, 1793

A collection of a hundred Great Brains makes one big fathead.

—Carl Jung, 1964

A committee can make a decision that is dumber than any of its members.

—Anonymous[3]

One man alone can be pretty dumb sometimes, but for real bona fide stupidity, there ain't nothing can beat teamwork.

—Edward Abbey, 1975

---

[2]A later paraphrase of this statement by W. Somerset Maugham (with the corresponding adjustment of the number :-) is also frequently quoted.

[3]In several collections, this wisdom is attributed to some David B. Coblitz, but I could not find a reliable confirmation of this authorship.

Insanity in individuals is something rare - but in groups, parties, nations and epochs, it is the rule.

—Friedrich Nietzsche, 1888

Men [...] go mad in herds, while they only recover their senses slowly, and one by one.

—Charles Mackay, 1841

When one goose drinks, all drink.

—German proverb

When people are free to do as we please, they usually imitate each other.

—Eric Hoffer, 1955

Where all men think alike, no one thinks very much.

—Walter Lippmann, 1915

You cannot make a man by standing a sheep on its hind-legs. But by standing a flock of sheep in that position you can make a crowd of men.

—Max Beerbohm, 1911

He who has earned no name,
Nor strives for noble things,
Belongs just to the elements...

—Johann Wolfgang von Goethe, 1831[4]

What men call social virtues, good fellowship, is commonly but the virtue of pigs in a litter, which lie close together to keep each other warm.

—Henry David Thoreau, 1854

In an absolutely corrupt age, such as the one we are living in, the safest course is to do as the others do.

—Marquis de Sade, 1795

Where ignorance is bliss, 'tis folly to be wise.

—Thomas Gray, 1742

***

To respect a sick age is to be in contempt of eternity.

—John Ciardi, 1962

It is no measure of health to be well-adjusted to a profoundly sick society.

—Jiddu Krishnamurti, 1975

He who has an opinion of his own, but depends on the opinion and tastes of others is a slave.

—Friedrich Gottlieb Klopstock, 1774

---

[4]From Part 2 of his *Faust*. (Translation by Walter Arndt).

Wherever there is authority, there is a natural inclination to disobedience.

—*Thomas Chandler Haliburton*, 1853

If they give you ruled paper, write the other way.[5]

—*Juan Ramón Jiménez*, 1953

A gentleman is always in the minority, it's his privilege.

—*Richard Aldington*, 1933

In a nation of sheep, one brave man forms a majority.

—*Edward Abbey*, 1990

Human salvation lies in the hands of the creatively maladjusted.

—*Martin Luther King Jr.*, 1963

Whoso would be a man must be a nonconformist.

—*Ralph Waldo Emerson*, 1841

In this age, the mere example of nonconformity, the mere refusal to bend the knee to custom, is itself a service.

—*John Stuart Mill*, 1859

There is a level of cowardice lower than that of the conformist. It is the fashionable non-conformist.

—*Ayn Rand*, 1971

To think is to differ.

—*Clarence Darrow*, 1925

When we lose the right to be different, we lose the privilege to be free.

—*Charles Evans Hughes*, 1925

A person does not have to be turned into a puppet jerked by social controls. The solution is to gradually become free of societal rewards and learn how to substitute for them rewards that are under one's own powers.

—*Mihaly Csikszentmihalyi*, 1990

We are more wicked together than separately. If you are forced to be in a crowd, then most of you should withdraw into your self.

—*Seneca*, ∼65 AD

The best things and best people rise of their separateness.

—*Robert Frost*[6]

***

---

[5]Ray Bradbury used this line as the epigraph for his famous *Fahrenheit 451*.

[6]As quoted, without a date, in many collections, starting not later than 1980.

Content thyself to be obscurely good.
When vice prevails, and impious men bear sway,
The post of honor is a private station.

—Joseph Addison, 1713

It is easy in the world to live after the world's opinion; it is easy in solitude to live after our own; but the great man is he who in the midst of the crowd keeps with perfect sweetness the independence of solitude.

—Ralph Waldo Emerson, 1840

I never found companion that was so companionable as solitude.

—Henry David Thoreau, 1854

What is necessary, after all, is only this: solitude, vast inner solitude.

—Rainer Maria Rilke, 1929

The world is a prison in which solitary confinement is preferable.

—Jiddu Krishnamurti, 1971[7]

Do not rely completely on any other human being, however dear. We meet all life's greatest tests alone.

—Agnes Macphail, 1924

The strongest man in the world is the man who stands alone.

—Henrik Ibsen, 1882

The heroes, the saints and sages – they are those who face the world alone.

—Norman Douglas, 1917

To be alone is the fate of all great minds – a fate deplored at times, but still always chosen as the less grievous of two evils.

—Arthur Schopenhauer, 1851

To be alone is to be different, to be different is to be alone.

—Suzanne Gordon, 1976

To be an adult is to be alone.

—Jean Rostand, 1954

All our evils come from not being able to be on our own.

—Jean de la Bruyère, 1688

If you are afraid of being lonely, don't try to be right.

—Jules Renard[8]

***

---

[7]Sometimes misattributed to Karl Kraus, who arranged the 1971 publication of J. Krishnamurti's work, and sometimes quoted him.

[8]As quoted, without a date, in a few generally reliable collections, starting not later than 1977.

Every man is like the company he loves to keep.

—Euripides, ~410 BC

Every man is known by the company he avoids.

—Anonymous[9]

Love your neighbor as thyself, but choose your neighborhood.

—Anonymous[10]

Love your neighbor, yet pull not down your hedge.

—George Herbert, 1640

The Bible tells us to love our neighbors, and also to love our enemies; probably because they are generally the same people.

—G. K. Chesterton, 1910

***

People with courage and character always seem sinister to the rest.

—Hermann Hesse, 1919

Excellence makes people nervous.

—Shana Alexander, 1970

Every thinker puts some portion of an apparently stable world in peril.

—John Dewey, 1929

Ours is the age that is proud of machines that think and suspicious of men who try to.

—Howard Mumford Jones, 1951

When a true genius appears in this world you may know him from this sign, that the dunces are all in confederacy against him.

—Jonathan Swift, 1706

Great spirits have always encountered violent opposition from mediocre minds.

—Albert Einstein, 1940

The artist and the multitude are natural enemies.

—Robert Altman, 1976

Great innovators and original thinkers and artists attract the wrath of mediocrities as lightning rods draw the flashes.

—Theodor Reik, 1963

---

[9]I have seen this twist of Euripides' aphorism in a book published as early as in 1915—without a reference.

[10]I have seen this extension of a Bible's commandment (*Leviticus* 19:18) attributed to some Louise Beal (could that be Louise Lester Beal, 1867-1952?), but could not find a reliable confirmation of this authorship.

In the land of the blind, the one-eyed man is king.

Erasmus, 1500[11]

In the land of the blind, the one-eyed man is stoned to death.

—*Joan D. Vinge*, 1998[12]

When an individual endeavors to lift himself above his fellows, he is dragged down by the mass, either by means of ridicule or of calamity.

—*Heinrich Heine*, 1830

The nail that sticks out gets hammered down.

—*Japanese proverb*

If you stand tall, you'll be shot at. If you stoop down, you'll be stepped on.

—*Jess Keating*, 2014[13]

When a man is down, everybody runs over him.

—*German proverb*

***

When people have no other tyrant, their own public opinion becomes one.

—*Edward Bulwer-Lytton*, 1872

You may talk about the tyranny of Nero and Tiberias, but the real tyranny is the tyranny of your next-door neighbor.

—*Walter Bagehot*, 1889

Tyranny and despotism can be exercised by many - more rigorously, and more severely, than by one.

—*Andrew Johnson*, 1866

Whatever crushes individuality is despotism, by whatever name it may be called.

—*John Stuart Mill*, 1859

The crowd plays the tyrant, when it is not in fear.

—*Baruch Spinoza*, 1677

Public is a ferocious beast: one must chain it up or flee from it.

—*Voltaire*, 1748

Public opinion, because of the tremendous urge to conformity in gregarious animals, is less tolerant than any system of laws.

—*George Orwell*, 1946

---

[11]This is a shorter paraphrase of a maxim from *Genesis Rabbah*, circa the 4th or 5th century BC.

[12]This aphorism may serve as a summary of a (wonderful) much earlier story by H. G. Wells.

[13]I have seen a similar quote attributed to some Carlos A. Urbizo, without a date, but could not confirm his authorship or even identify this person.

When we lose our individual independence in the corporateness of a mass movement, we find a new freedom – freedom to hate, bully, lie, torture, murder, and betray without shame and remorse.

—*Eric Hoffer*, 1980

***

Every numerous assembly is *mob*, let the individuals who compose it be what they will.

—*Lord Chesterfield*, 1751

A mob is a society of bodies voluntarily bereaving themselves of reason.

—*Ralph Waldo Emerson*, 1841

There's a whiff of the lynch mob or the lemming migration about any overlarge concentration of like-thinking individuals, no matter how virtuous their cause.

—*P. J. O'Rourke*, 1991

Much of America's intelligentsia has become a mob.

—*George Will*, 2020

I want men to be free as much from mobs as kings.

—*Lord Byron*, 1821

***

Evil draws men together.

—*Aristotle*[14]

I am a member of a party of one, and I live in an age of fear.

—*E. B. White*, 1947

Woods have ears and fields have eyes.

—*European proverb*

There is no defense against reproach but obscurity.

—*Joseph Addison*, 1711

In an expanding universe, time is on the side of an outcast.

—*Quentin Crisp*, 1968

An effective way to deal with predators is to taste terrible.

—*Anonymous*[15]

There is no safety in numbers, or in anything else.

—*James Thurber*, 1939[16]

---

[14]From his *Rhetoric* (written in 4th century BC).

[15]This maxim is frequently misattributed to A. Blanchard et al., who used it in the title of their 2016 paper, but I have seen it published earlier as an anonymous joke.

[16]The expression "safety in numbers" may be traced back at least to a Latin saying, "*Defendit Numerus*".

Sometimes I get the feeling the whole world is against me, but deep down I know that's not true. Some smaller countries are neutral.

—Robert Orben, 2008

Men are not against you; they are merely for themselves.

—Gene Fowler, 1961

***

Give me the liberty to know, to utter, and to argue freely according to my conscience, above all liberties.

—John Milton, 1644

Liberty [...] is the sovereignty of the individual.

—Josiah Warren, 1846

Individuality is the aim of political liberty.

—James Fenimore Cooper, 1848

The right to be left alone is indeed the beginning of all freedom.

—William O. Douglas, 1952[17]

A free society is a society where it is safe to be unpopular.

—Adlai E. Stevenson, 1952

[Freedom from fear] may be said to sum up the whole philosophy of human rights.

—Dag Hammarskjöld, 1956

If liberty means anything at all, it means the right to tell people what they to not want to hear.

—George Orwell, 1972

***

Liberty consists in the power to do anything that does not injure others.

—French National Assembly, 1789

Your right to swing your fist ends where my nose begins.

—Anonymous[18]

I detest what you write, but I would give my life to make it possible for you to continue to write.

—Voltaire, 1770[19]

---

[17]From his dissenting opinion in the US Supreme Court.

[18]This aphorism, based on a longer joke by John B. Finch (circa 1882) is sometimes misattributed to Abraham Lincoln, Oliver Wendell Holmes, Jr., and John Stewart Mill.

[19]A more popular, later (circa 1906) paraphrase of this sentence, "I disapprove of what you say, but will defend to the death your right to say so", belongs to Evelyn Beatrice Hall (under the pen name Stephen G. Tallentyre).

Liberty, as well as honor, man ought to preserve at the hazard of his life, for without it, life is unsupportable.

—Miguel de Cervantes, 1615

Freedom cannot be bought for nothing. If you hold it precious, you must hold all else of little value.

—Seneca, ∼ 65 AD

Those who would give up essential Liberty, to purchase a little temporary Safety, deserve neither Liberty nor Safety.

—Benjamin Franklin, 1755

If a nation values anything more than freedom, it will loose its freedom.

—W. Somerset Maugham, 1941

The condition upon which God hath given liberty to man is eternal vigilance.

—John Philpot Curran, 1790[20]

He only earns his freedom and existence
Who daily conquers them anew.

—Johann Wolfgang von Goethe, 1832[21]

God grants liberty only to those who love it and are always ready to guard and defend it.

—Daniel Webster, 1834

There is only one basic human right, the right to do as you damn well please. And with it comes the only basic human duty, the duty to take the consequences.

—P. J. O'Rourke, 1993

***

Liberty means responsibility. That is why most men dread it.

—George Bernard Shaw, 1903

A hero as someone who understands the degree of responsibility that comes with his freedom.

—Bob Dylan, 1985

Man is tormented by no greater anxiety than to find someone quickly to whom he can hand over that great gift of freedom with which the ill-fated creature is born.

—Fyodor Dostoevsky, 1880

---

[20]A shorter paraphrase of this statement, "The price for freedom is eternal vigilance," is frequently attributed to Thomas Jefferson (starting from 1834), but I am not aware of any reliable confirmation of his authorship.

[21]From Part 2 of his *Faust*. (Translation by Walter Arndt.)

The average man's love of liberty is nine-tenth imaginary, exactly like his love of sense, justice, and truth.

—H. L. Mencken, 1923

In our country we have those three unspeakably precious things: freedom of speech, freedom of conscience, and the prudence never to practice either of them.

—Mark Twain, 1897

He was jeopardizing his traditional rights of freedom and independence by daring to exercise them.

—Joseph Heller, 1961

***

Since people will never cease trying to interfere with the liberties of others in pursuing their own, the State can never wither away.

—W. H. Auden, 1946

Society is produced by our wants and government by our wickedness.

—Thomas Paine, 1776

If men were angels, no government would be necessary.

—James Madison, 1788

Why has government been instituted at all? Because the passions of men will not conform to the dictates of reason and justice, without constraint.

—Alexander Hamilton, 1788

The only use of government is to repress the vices of man.[22]

—Percy Bysshe Shelley, 1812

It may be true that the law cannot make a man love me, but it can stop him from lynching me, and I think that's pretty important.

—Martin Luther King Jr., 1964

Those who fear men like laws.

—Luc de Clapiers, 1746

The most desirable state of mankind, is that which maintains general security, with the smallest encroachment upon individual independence.

—William Godwin, 1793

***

The most powerful factors in the world are clear ideas in the minds of energetic men of good will.

—Arthur Thomson, 1922

---

[22]For more on government, see the next section.

Sentiment without action is the ruin of the soul.

—Edward Abbey, 1990

Action is but coarsened thought.

—Henri-Frédéric Amiel, 1850

Action is [...] the enemy of thought and the friend of flattering illusions.

—Joseph Conrad, 1904

The people I distrust most are those who want to improve our lives but have only one course of action in mind.

—Frank Herbert, 1981

God save me from the man of one book.

—Italian proverb

The road to hell is paved with good intentions.

—European proverb[23]

Human beings are perhaps never more frightening than when they are convinced beyond doubt that they are right.

—Laurens van der Post, 1958

The greatest dangers to liberty lurk in insidious encroachment by men of zeal, well meaning but without understanding.

—Louis D. Brandeis, 1928

Convictions are more dangerous enemies of truth than lies.

—Friedrich Nietzsche, 1878

Nearly all our disasters come of a few fools having the 'courage of their convictions'.

—Coventry Patmore, 1895

One person with a belief is a social power equal to ninety-nine who have only interests.

—John Stuart Mill, 1861

A blind man leads a crowd.

—Japanese proverb

An invasion of armies can be resisted; an invasion of ideas cannot be resisted.

—Victor Hugo, 1852[24]

Man is ready to die for an idea, provided that this idea is not very clear for him.

—Paul Eldridge, 1963

---

[23]This proverb, frequently misattributed to Samuel Johnson, is apparently based on the 1605 maxim "Hell is full of good intentions or desires" by Francis de Sales.

[24]Numerous later paraphrases of this line frequently included the qualifier "the ideas whose time has come".

To be willing to die for an idea is to set a rather high price upon a conjecture.

—Anatole France, 1914

I would never die for my beliefs because I might be wrong.

—Bertrand Russell, 1961

The main dangers in this life are the people who want to change everything - or nothing.

—Nancy Astor[25]

To go beyond is as wrong as to fall short.

—Confucius, fifth century BC

All movements go too far.

—Bertrand Russell, 1950

I feel uneasy at the very idea of a Movement. I see every insight degenerating into a dogma, and fresh thoughts freezing into lifeless party line.

—I. F. Stone, 1969

There is no belief, however foolish, that will not gather its faithful adherents who will defend it to the death.

—Isaac Asimov, 1974

The more fantastic an ideology or theology, the more fanatic its adherents.

—Edward Abbey, 1990

Fanaticism consists in redoubling your effort when you have forgotten your aim.

—George Santayana, 1905

***

Usually, terrible things that are done under the excuse that progress requires them are not really the progress at all, but just terrible things.

—Russell Baker, 1970

What we call 'Progress' is the exchange of one Nuisance for another Nuisance.[26]

—Havelock Ellis, 1914

What the world turns to, when it has been cured of one error, is usually simply another error, and maybe one worse than the first one.

—H. L. Mencken, 1922

---

[25]As quoted, without a date, by many, including several reliable sources, starting not later than 1992.

[26]This aphorism (just as the next three quotes) is essentially a twist of the famous Goethe's characterization (in Part 2 of his *Faust*) of the perennial "struggles for liberation" as the replacements of one enslavement with another enslavement. (Sorry, I could not find a fair poetic English translation of that verse).

Revolution, *n*. In politics, an abrupt change in the form of misgovernment.

—*Ambrose Bierce*, 1911

Revolutions have never lightened the burden of tyranny: they only shifted it to another shoulder.

—*George Bernard Shaw*, 1903

Revolution is a trivial shift in the emphasis of suffering; the capacity for self-indulgence changes hands.

—*Tom Stoppard*, 1966

Revolution evaporates, and leaves behind only the slime of a new bureaucracy.

—*Franz Kafka*, 1920

It is doubtful if the oppressed ever fight for freedom. They fight for pride and power – power to oppress others.

—*Eric Hoffer*, 1951

The urge to save humanity is almost always only a false-face for the urge to rule it. Power is what all messiahs really seek: not the chance to serve.

—*H. L. Mencken*, 1949

Constant experience shows us that every man invested with power is apt to abuse it.

—*Montesquieu*, 1748

Watch out for the fellow who talks about putting things in order! Putting things in order always means getting other people under your control.

—*Denis Diderot*, 1796

The psychopaths are always around. In calm times we study them, but in times of upheaval, they rule over us.

—*Ernst Kretschner*, 1963

He who fights with monsters should be careful to not become one.

—*Friedrich Nietzsche*, 1886

Every revolutionary ends up as an oppressor or a heretic.

—*Albert Camus*, 1951

The revolution, like Saturn, will successively devour its children, and finally produce despotism.

—*Pierre Vergniaud*, 1793[27]

Whether a revolution succeeds or miscarries, men of great hearts will always be the victims.

—*Heinrich Heine*, 1834

The unselfish and the intelligent may begin a movement – but it passes away from them. They are not the leaders of the revolution. They are its victims.

—*Joseph Conrad*, 1911

---

[27]This prediction was confirmed, in just a few months, by author's own fate.

Saints are usually killed by their own people.

—Eric Sevareid, 1968

The most radical revolutionary will become a conservative the day after the revolution.

—Hannah Arendt, 1970

Not to be a republican at twenty is proof of want of heart; to be one at thirty is proof of want of head.

—François Guizot, mid-1800 s[28]

I never dared to be radical when young
For fear it would make me conservative when old.

—Robert Frost, 1936

***

Most of the change we think we see in life
Is due to truths being in and out of favor.

—Robert Frost, 1914

If it is not necessary to change, it is necessary not to change.

—Lucius Cary, 2nd Viscount Falkland, 1641

All human situations have their inconveniences. We feel those of the present but neither see nor feel those of the future, and hence we often make troublesome changes without amendment, and frequently for the worse.

—Benjamin Franklin, 1780

The art of progress is to preserve order amid change, and to preserve change amid order.

—Alfred North Whitehead, 1929

From barbarism to civilization requires a century; from civilization to barbarism just a day.

—Will Durant, 1957

A society can be progressive only if it conserves its traditions.

—Walter Lippmann, 1955

***

Everyone thinks of changing humanity, but nobody thinks of changing himself.

—Leo Tolstoy, 1900

I never expect to see a perfect work from imperfect man.

—Alexander Hamilton, 1788

---

[28]Later this statement was rephrased, among others, by George Bernard Shaw, Benjamin Disraeli, and Otto von Bismarck. Perhaps the most popular of its forms is "If you are not a revolutionary at twenty, you are a rascal; if you are not a conservative at thirty, you are a fool".

The belief, not only of Socialists but also of those so-called Liberals who are diligently preparing the way for them, is that by due skill an ill-working humanity may be framed in well-working institutions. It is a delusion.

—*Herbert Spencer*, 1884

You cannot hope to build a better world without improving the individuals.

—*Marie Curie*, 1923

The important work of moving the world forward does not wait to be done by perfect men.

—*George Eliot*, 1858

The reasonable man adapts himself to the world; the unreasonable one persists in trying to adapt the world to himself. Therefore all progress depends on the unreasonable man.[29]

—*George Bernard Shaw*, 1903

***

Adapt or perish, now as ever, is Nature's inexorable imperative.

—*H. G. Wells*, 1945

Natural selection determined the evolution of cultures in the same manner as it did that of species.

—*Konrad Lorenz*, 1966

The history of man is a graveyard of great cultures that came to catastrophic ends because of their incapacity for planned, rational, voluntary reaction to challenge.

—*Erich Fromm*, 1962

We civilizations now know ourselves mortal.

—*Paul Valéry*, 1919

The end of the human race will be that it will eventually die of civilization.

—*Ralph Waldo Emerson*, 1870

The chief obstacle to the progress of the human race is the human race.

—*Don Marquis*, 1927

The people that once bestowed commands, consulships, legions, and all else, now meddles no more, and longs eagerly for just two things - bread and circuses!

—*Juvenal*, 2nd century AD

We used to build civilizations. Now we build shopping malls.

—*Bill Bryson*, 1991[30]

Civilizations die by suicide, not by murder.

—*Arnold J. Toynbee*, 1947

---

[29]A popular (anonymous?) extension of this quote is: "...this becomes evident if you watch that 'progress' for a while".

[30]Note the date; by now we have stopped doing even that – besides those in computer games.

Civilization begins with order, grows with liberty, and dies with chaos.

—Will Durant, 1926

Expansion means complexity, and complexity decay.

—C. Northcote Parkinson, 1962

To accept civilization as it is practically means accepting decay.

—George Orwell, 1968

Civilization, like an airplane in flight, survives only if it keeps going forward.

—Edward Abbey, 1990

We need to be bold and adventurous in our thinking to survive.

—William O. Douglas, 1952

On Government and Politics

Government was intended to suppress injustice, but its effect has been to embody and perpetuate it.

—William Godwin, 1793

Government exists to protect us from each other. Where government has gone beyond its limits is in deciding to protect us from ourselves.

—Ronald Reagan, 1980[1]

The government solution to a problem is usually as bad as the problem and very often makes the problem worse.

—Milton Friedman, 1975

If we can prevent the government from wasting the labors of the people, under the pretense of taking care of them, they must become happy.

—Thomas Jefferson, 1802

The freest form of government is only the least objectionable form.

—Herbert Spencer, 1851

Anarchism is founded on the observation that since few men are wise enough to rule themselves, even fewer are wise enough to rule others.

—Edward Abbey, 1990

The worst thing in the world next to anarchy, is government.

—Henry Ward Beecher, 1867

***

---

[1]I wish I knew the name of his speechwriter.

K. K. Likharev (ed.), *Essential Quotes for Scientists and Engineers*,
https://doi.org/10.1007/978-3-030-63332-5_18

The world is run by 'C' students.[2]

—*Al McGuire*[3]

Decisions are made by people who have time, not people who have talent.

—*Scott Adams*, 2011

We hang the petty thieves and appoint the great ones to public office.

—*Aesop*[4]

Under our republican form of government, intelligence is so highly honored that it is rewarded by exemption from the cares of office.

—*Ambrose Bierce*, 1906

I am strongly in favor of common sense, common honesty, and common decency. This makes me forever ineligible to any public office.

—*H. L. Mencken*, 1946

***

There is a tragic flaw in our precious Constitution, and I don't know what can be done to fix it. This is it: Only nut cases want to be president.

—*Kurt Vonnegut*, 2004

Anybody that wants the presidency so much that he'll spend two years organizing and campaigning for it is not to be trusted with the office.

—*David S. Broder*, 1973[5]

When I was a boy I was told that anybody could become President. I'm beginning to believe it.

—*Clarence Darrow*, 1941

On some great and glorious day the plain folks of the land will reach their heart's desire at last, and the White House will be adorned by a downright moron.

—*H. L. Mencken*, 1920[6]

***

There are men, in all ages, who [...] mean to govern well, but they mean to govern. They promise to be kind masters, but they mean to be masters.

—*Daniel Webster*, 1837

---

[2] If you have any doubt, just watch the news—or apply for a research grant.

[3] As quoted, without a date, in several collections, starting not later than 1987.

[4] As quoted, without a date, in major collections of his fables.

[5] A later (circa 2005) paraphrase of this maxim by Douglas Adams is quoted even more frequently.

[6] Note the date. If written today, the verbs' tense might be different....

Power tends to corrupt, and absolute power corrupts absolutely. Great men are almost always bad men.

—Lord Acton, 1887[7]

It is said that power 'corrupts', but actually it's more true that power *attracts the corruptible*. The sane are usually attracted by other things than power.

—David Brin, 1985

All governments suffer a recurring problem: Power attracts pathological personalities. It is not that power corrupts but that it is magnetic to the corruptible.

—Frank Herbert, 1985

Power dements even more than it corrupts, lowering the guard of foresight and raising the haste of action.

—Will and Ariel Durant, 1975

Authority has always attracted the lowest elements in the human race. All through history mankind has been bullied by scum.

—P. J. O'Rourke, 1991

The higher the ape climbs, the more he shows his naked rump.

—European proverb

No man can rule innocently.

—Louis Antonie de Saint-Just, 1792

***

The great mass of men are made by nature to be slaves, they are unfit to control themselves, and for their own good need masters.

—W. Somerset Maugham, 1939

Nor should we listen to those who say, 'The voice of the people is the voice of God', for the turbulence of the mob is always close to insanity.

—Alcuin, 800 AD

Mankind, when left to themselves, are unfit for their own government.

—George Washington, 1786[8]

The proposition, that [the people] are the best keepers of their own liberties is not true. They are the worst conceivable; they are no keepers at all. They can neither judge, act, think, or will, as a [political body].

—John Adams, 1787[9]

---

[7]This famous line is actually a shorter paraphrase of an earlier (circa 1770) aphorism by William Pitt. In turn, it was repeatedly twisted later, including the notorious maxim "Absolute liberality corrupts absolutely" by Gertrude Himmelfarb. (See also the next three quotes).

[8]I still remember my shock at first reading this statement—by *this* author.

[9]One more quote in the same spirit—from another Founding Father of the United States – also from his private correspondence of course.

The masses [...] neither should nor can direct their personal existence, and still less to rule society in general.

—José Ortega y Gasset, 1930

The ignorant classes are the dangerous classes. Ignorance is the womb of monsters.

—Henry Ward Beecher, 1867

In the past, civilizations have been created and directed by a small intellectual aristocracy, never by the crowd.

—Gustave Le Bon, 1895

There are no wise few. Every aristocracy that has ever existed has behaved, in all essential points, exactly like a small mob.

—G. K. Chesterton, 1905

The intellectual elite is the most heavily indoctrinated sector, for good reasons. It's their role as a secular priesthood to really believe the nonsense that they put forth.

—Noam Chomsky, 1987

Plutocracy is abhorrent to a republic; it is more despotic than monarchy, more heartless than aristocracy, more selfish than bureaucracy.

—William Jennings Bryan, 1906

The danger is not that a particular class is unfit to govern. Every class is unfit to govern.

—Lord Acton, 1881

The best government is a benevolent tyranny tempered by an occasional assassination.

—Voltaire[10]

The best system of government would be an enlightened monarchy, but it has the unsolved problem of uninterrupted supply of enlightened monarchs.

—Anonymous

In the long run every Government is the exact symbol of its People, with their wisdom and unwisdom; we have to say, Like People like Government.

—Thomas Carlyle, 1843

Who serves the people, has a bad master.

—European proverb

***

Anyone can hold the helm when the sea is calm.

—Publilius Syrus, first century BC

---

[10]As quoted, without a date, by Laurence J. Peter in 1977.

You may proclaim, dear sirs, your fine philosophy,
But till you feed us, right and wrong can wait!

—Bertolt Brecht, 1928

The horses fight when the manger is empty.

—Danish proverb

[Karl Marx] discovered the simple fact that, hitherto concealed by an overgrowth of ideology, that mankind must first eat and drink, have shelter and closing, before it can pursue politics, science, religion, art, etc.

—Friedrich Engels, 1883

The important thing is the struggle everybody is engaged into get better living conditions, and they are not interested too much in the form of government.

—Bernard Baruch, 1964

Few men desire liberty; most wish only for a just master.

—Sallust, 1st century BC

When men are well governed, they neither seek nor desire any other liberty.

—Niccoló Machiavelli, 1517

***

Pity the meek, for they shall inherit the earth.

—Don Marquis[11]

The century on which we are entering – the century which we will come out of this war – can be and must be the century of the common man.

—Henry A. Wallace, 1942

We are in danger of developing a cult of the Common Man, which means a cult of mediocrity.

—Herbert Hoover, 1955

A government by the passions of the multitude, or, no less correctly, according to the vices and ambition of their leaders, is a democracy.

—Fisher Ames, 1805

What does the democracy come down to? The persuasive power of slogans invented by wily self-seeking politicians.

—W. Somerset Maugham, 1939

Democracy is a pathetic belief in the collective wisdom of individual ignorance.

—H. L. Mencken, 1926[12]

---

[11]This is of course an (undated) twist of a line from the Bible (Matthew 5:5), with the word "pity" replacing the original "blessed".

[12]*Wikiquotes* states that this is a misattribution, but it is not – a rare error on this great web site.

It has been observed that a pure democracy, if it were practicable, would be the most perfect government. Experience has proved that no position in politics is more false than this.

—*Alexander Hamilton*, 1788

The best argument against democracy is a five-minute conversation with the average voter.

—*Anonymous*[13]

Democracy is the worst system of government. It reduces wisdom to impotence and secures the triumph of folly, ignorance, clap-trap and demagogy. […] Yet democracy is the only form of social order that is admissible, because it is the only one consistent with justice.

—*Robert Briffault*, 1930

No one pretends that democracy is perfect or all-wise. Indeed, it has been said that democracy is the worst form of government except all the others that have been tried.

—*Winston Churchill*, 1947

In a democracy the majority of citizens is capable of exercising the most cruel oppression upon the minority.

—*Edmund Burke*, 1790

It is bad to be oppressed by a minority, but it is worse to be oppressed by a majority.

—*Lord Acton*, 1877

***

The love of democracy is that of equality.[14]

—*Montesquieu*, 1748

'All animals are equal, but some are more equal than others.'

—*George Orwell*, 1945

Democratic institutions awaken and foster a passion for equality which they can never satisfy.

—*Alexis de Tocqueville*, 1893

It is not true that equality is a law of nature. Nature has made nothing equal.

—*Luc de Clapiers*, 1746

That all men are equal is a proposition to which, at ordinary times, no sane human being has ever given his assent.

—*Aldous Huxley*, 1927

A society that puts equality ahead of freedom will end up with neither.

—*Milton Friedman*, 1980

---

[13]Frequently misattributed to Winston Churchill. For his actual quote on this subject, see below.

[14]For me, as a scientist trained to discuss only clearly defined notions, it was always amazing to see how some authors lacking such training confuse—whether intentionally or unintentionally—two radically different notions of equality: that of *human rights*, and the equality (or inequality) of *personal benefits*—income, health care, housing—you name it. This confusion allows them to argue ad infinitum, without any risk of reaching an agreement—see the quotes below.

Where freedom is real, equality is the passion of the masses.
Where equality is real, freedom is the passion of a small minority.

—Eric Hoffer, 1951

***

Conservative, *n.* A statesman enamored of existing evils, as distinguished from a Liberal, who wants to replace them with others.

—Ambrose Bierce, 1906

Counterpart to the knee-jerk liberal is a knee-pad conservative.

—Edward Abbey, 1990

Conservative government is an organized hypocrisy.[15]

—Benjamin Disraeli, 1845

The world is full of people whose notion of a satisfactory future is, in fact, a return to the idealized past.

—Robertson Davies, 1960

In every age 'the good old days' were a myth. [...] For every age has consisted of crises that seemed intolerable to the people who lived through them.

—Brooks Atkinson, 1951

He that will nor apply new remedies must expect new evils; for the time is the greatest innovator.

—Francis Bacon, 1625

A state without the means of some change is without the means of its conservation.

—Edmund Burke, 1790

A tradition is kept alive only by something being added to it.

—Henry James, 1888

The past is for inspiration, not imitation, for continuation, not repetition.

—Israel Zangwill, 1921

Because we don't think about future generations, they will never forget us.

—Henrik Tikkanen[16]

***

What experience and history teach is this – that nations and governments have never learned anything from history.

—Georg Wilhelm Friedrich Hegel, 1832[17]

---

[15]Could this mean that the liberal government's hypocrisy is always disorganized?

[16]As quoted, without a date, in a few generally reliable sources, starting not later than 2005.

[17]This maxim is often misattributed to George Bernard Shaw who actually used it with a reference to Hegel, and also to Aldous Huxley who just paraphrased it.

Those who cannot remember the past are condemned to repeat it.

—George Santayana, 1905[18]

The past is never dead. It's not even past.

—William Faulkner, 1951

I have seen the future and it doesn't work.

—Robert Fulford[19]

It would only take one generation of forgetfulness to put us back intellectually several hundred years.

—Anonymous[20]

When a nation goes down, or a society perishes, one condition may always be found. They forgot where they came from. They lost sight of what had brought them along.

—Carl Sandburg, 1948

***

Liberalism has no obvious answers to the biggest problems we face: ecological collapse and technological disruption.

—Yuval Noah Harari, 2018

A liberal is a man who is willing to spend somebody else's money.

—Carter Glass[21]

A liberal is a person whose interests aren't at stake at the moment.

—Willis Player, 1972

Much of the social history of the Western world over the past three decades has involved replacing what worked with what sounded good.

—Thomas Sowell, 1993

There's a mighty big difference between good, sound reasons and reasons that sound good.

—William E. Vaughan, 1978

Tender surgeons make foul wounds.

—European proverb

Nothing emboldens sin so much as mercy.

—William Shakespeare, 1606

---

[18]This famous aphorism is actually a succinct paraphrase of earlier, longer lines by Baruch Spinoza and Edmund Burke.

[19]This is an evident reverse of the well-known statement made by Lincoln Steffens after his visit to Soviet Russia in 1921, or rather to its later (circa 1933) paraphrase: "I have seen the future and it works   ".

[20]This statement is sometimes attributed to certain Dean Tollefson, but I could not find a reliable confirmation of this authorship.

[21]As quoted, without a date, in several reputable collections.

He who spares vice wrongs virtue.

—European proverb

Pity for the guilty is treason to the innocent.

—Ayn Rand, 1969

Whenever the government assumes to deliver us from the trouble of thinking for ourselves, the only consequence it produces are those of torpor and imbecility.

—William Godwin, 1793

The weak members of civilized society propagate their kind. [...] This may be highly injurious to the race of men.

—Charles Darwin, 1871

Sympathy thwarts the law of development, of evolution, of the survival of the fittest. It preserves what is ripe for extinction.

—Friederich Nietzsche, 1888

The ultimate result of shielding men from the effects of their folly, is to fill the world with fools.

—Herbert Spencer, 1891

A nation or civilization that continues to produce soft-minded men purchases its own spiritual death on an installment plan.

—Martin Luther King Jr., 1963

***

One of the serious obstacles to the improvement of our race is indiscriminate charity.

—Andrew Carnegie, 1889[22]

Charity corrupts both the giver and the taker, and moreover does not reach its goal, because it merely multiplies poverty.

—Fyodor Dostoevsky, 1880

Charity is twice cursed – it hardens him that gives and softens him that takes.

—Bouck White, 1911

If you give a man a fish he is hungry again in an hour. If you teach him to catch a fish you do him a good turn.[23]

—Anne Isabella Thackeray Ritchie, 1885

---

[22]He was not some old scrooge, but perhaps the most famous American philanthropist, who also said, "The man who dies thus rich dies disgraced". The word "indiscriminate" is certainly the key to this apparent contradiction.

[23]This is actually a rather liberal translation of an old Chinese proverb; I quote it in this form only because it has served as the basis of numerous later twists. Of them, I especially like the following (anonymous?) wisdom: "If you promise a man somebody else's fish, he will vote for you".

The best way of doing good to the poor, is not making them easy in poverty, but leading or driving them out of it.

—Benjamin Franklin, 1766

***

You cannot bring about prosperity by discouraging thrift.
You cannot strengthen the weak by weakening the strong.
You cannot help little men by tearing down big men.
You cannot lift the wage earner by pulling down the wage payer.
You cannot help the poor by destroying the rich.
You cannot establish sound security on borrowed money.
You cannot further the brotherhood of man by inciting class hatred.
You cannot keep out of trouble by spending more than you earn.
You cannot build character and courage by taking away men's initiative and independence.
And you cannot help men permanently by doing for them what they could and should do for themselves.

—William J. H. Boetcker, 1916[24]

The national budget must be balanced. The public debt must be reduced [...] Payments to foreign governments must be reduced, if the nation doesn 't want to go bankrupt. People must again learn to work, instead of living on public assistance.

—Cicero, 55 BC[25]

Nothing is so well calculated to produce a death-like torpor in the country as an extended system of taxation and a great national debt.

—William Cobbett, 1804

Fundamentally, there are two ways of coordinating the economic activity of millions. One is central direction involving the use of coercion: the technique of the army and the modern totalitarian state. The other is voluntary cooperation of individuals: the technique of the marketplace.

—Milton Friedman, 1962

America has believed that in differentiation, not in uniformity, lies the path of progress. It acted on this belief; it has advanced human happiness, and it has prospered.

—Louis D. Brandeis, 1915

It is not from the benevolence of the butcher, the brewer, or the baker, that we expect our dinner, but from their regard of their own interest.

—Adam Smith, 1776

The worst crime against working people is a company which fails to operate at a profit.

—Samuel Gompers[26]

---

[24]These so-called *Ten Cannots* are often misattributed to Abraham Lincoln.

[25]Note the date; it takes millennia for some wisdoms to sink in ....

[26]As quoted by Michael Rothschild in 1990.

I like business because it is honestly selfish, thereby avoiding the hypocrisy and sentimentality of the unselfish attitude.

—William Feather, 1927

Sentimentality is the emotional promiscuity of those who have no sentiment.

—Norman Mailer, 1966

When we divorce capital from labor, capital is hoarded, and labor starves.

—Daniel Webster, 1838

Extinguish free enterprise and you extinguish liberty.

—Margaret Thatcher, 1979

Capitalism: Nothing so mean could be right.

—Edward Abbey, 1990

Good medicine tastes bitter.

—Chinese proverb

Capitalism, wisely managed, can probably be made more efficient for attaining economic ends than any alternative system yet in sight, but [...] in itself is extremely objectionable.

—John Maynard Keynes, 1926

The inherent vice of Capitalism is the unequal sharing of blessings; the inherent virtue of Socialism is the equal sharing of miseries.

—Winston Churchill, 1945

To grasp the true meaning of socialism, imagine a world where everything is designed by the post office, even the sleaze.

—P. J. O'Rourke, 1989

Socialism proposes no adequate substitute for the motive of enlightened selfishness that today is the basis of all human labor and effort, enterprise and new activity.

—William Howard Taft, 1913

When everybody own something, nobody owns it, and nobody has a direct interest in maintaining or improving its condition.

—Milton Friedman, 1980

What is everybody's business is nobody's business.

—European proverb

Socialism of any type and shade leads to a total destruction of the human spirit and to a leveling the mankind to death.

—Alexander Solzhenitsyn, 1978

Socialism in general has a record of failure so blatant that only an intellectual could ignore or evade it.

—Thomas Sowell, 1997

***

Socialism is Communism in course of construction; it is incomplete Communism.

—Nikolai Bukharin, 1919

From each according to his abilities, to each according to his needs.

—Karl Marx, 1875[27]

Communism is like prohibition, it's a good idea but it won't work.

—Will Rogers, 1927

Communism is the most painful path from capitalism to capitalism.

—Scott Adams, 1989[28]

***

Our rulers ought to be changed routinely, like diapers, for the same reason.

—Dick Nolan, 1966

Most people vote against somebody rather than for somebody.

—Franklin P. Adams, 1944

Vote for the man who promises least; he'll be least disappointing.

—Bernard Baruch, 1960

Under democracy one party always devotes its chief energies to trying to prove that the other party is unfit to rule - and both commonly succeed.

—H. L. Mencken, 1956

No one party can fool all of the people all of the time; that's why we have two parties.

—Bob Hope[29]

Each party is worse than the other. The one that's out always looks the best.

—Will Rogers, 1924

The Democrats are the party that says government will make you smarter, taller, richer, and remove crabgrass on your lawn. Republicans are the party that says government doesn't work, and then get elected and prove it.

—P. J. O'Rourke, 1991

Party is the madness of many, for the gain of a few.

—Alexander Pope, 1714

I am a lover of liberty. I will not and cannot serve a party.

—Erasmus, 1523

***

---

[27]This famous "sound-good" principle, based on earlier formulas by Henri de Saint-Simon (circa 1822) and Louis Blanc (circa 1841), was accepted by most communists as their final goal.

[28]Note the date.

[29]He was a comedian, and perhaps thought this observation was just a joke. (Sorry I could not establish its date).

Thanks to TV [...], you can only be one of two kinds of human beings, either a liberal or a conservative.

—Kurt Vonnegut, 2004[30]

People are so conditioned to take sides that a balanced analysis looks to them like hatred.

—Scott Adams, 2005

In politics [...] a community of hatred is almost always the foundation of friendship.

—Alexis de Tocqueville, 1893

There is no stronger bond of friendship than a common enemy.

—Frank Frankfort Moore, 1907

Love, friendship, respect do not unite people as much as a common hatred for something.

—Anton Chekhov, 1921

We hate some persons because we do not know them; and we will not know them because we hate them.

—Charles Caleb Colton, 1825

Listening to both sides of a story will convince you that there is more to a story than both sides.

—Frank Tyger, 1975

The middle of the road is all of the usable surface. The extremes, right and left, are in the gutters.

—Dwight Eisenhower, 1949

We know what happens to people who stay in the middle of the road. They get run over.

—Aneurin Bevan, 1953[31]

Between two stools one cannot sit, and falls to the ground.

—John Gower, 1390

There is no larger mistake than to try to leap an abyss in two jumps.

—David Lloyd George, 1933

United we stand, divided we fall.

—Aesop[32]

There are two kinds of fools: one says, 'This is old, therefore it is good'; the other says, 'This is new, therefore it is better'.

—Dean Inge, 1931

---

[30]Note the date. Nowadays, thanks to the Internet, you can only be (in the C. Northcote Parkinson's terminology) either an *in-law* or an *outlaw*.

[31]Possibly based on an earlier joke by Will Rogers.

[32]Somebody has noticed that at the current political polarization, this famous old statement seems instrumental only in bra advertisements.

God, give us grace to accept with serenity the things that cannot be changed, courage to change the things which should be changed, and the wisdom to distinguish the one from the other.

—Reinhold Niebuhr, 1942

∗∗∗

State is the kind of organization which, though it does big things badly, does small things badly, too.

—John Kenneth Galbraith, 1969

In a bureaucratic system, useless work drives out useful work.

—Milton Friedman, 1977

Work expands so as to fill the time available for its completion.

—C. Northcote Parkinson, 1955[33]

In a hierarchy every employee tends to rise to his level of incompetence. [...] Work is achieved by those employees who have not yet reached [that level].[34]

—Laurence J. Peter, 1969

If the federal government had improved in efficiency as much as the computer has since the 1950s, we'd only need four federal employees and the federal budget would be $100,000.

—Newt Gingrich, 1994[35]

The only thing that saves us from the bureaucracy is inefficiency. An efficient bureaucracy is the greatest threat to liberty.

—Eugene McCarthy, 1979

Whenever you have an efficient government you have a dictatorship.

—Harry S. Truman, 1959

Be thankful we're not getting all the government we're paying for.

—Will Rogers[36]

For every action there is an equal and opposite government program.

—Bob Wells[37]

---

[33]This is the punchline of his brilliant book *Parkinson's Law*, which summarizes the author's experience in the British Civil Service, and is perhaps be the best description of the basic features of all bureaucracies—including the scientific ones. Indeed, one of his later (circa 1962) twists of the law is: "Successful research attracts the bigger grant which makes further research impossible".

[34]This is the so-called *Peter Principle*, the main idea of another wonderful, very funny (but also realistic to the point of being scary) book with the same title, which looks at bureaucracies from a different angle than *Parkinson's Law*.

[35]Cf. the Robert X. Cringley's quote in the *On Science and Technology* section.

[36]Included, without a date, into the 1997 collection of his quotes.

[37]As quoted, without a date, by many, starting not later than 1998.

Governments never learn.

—Milton Friedman, 1980

There 's no trick to being a humorist when you have the whole government working for you.

—Will Rogers[38]

***

Government expands to absorb revenue - and then some.[39]

—Tom Wicker, 1964

It is easy to be generous with another man's purse.

—European proverb

The first lesson of economics is scarcity: there is never enough of anything to fully satisfy all those who want it. The first lesson of politics is to disregard the first lesson of economics.

—Thomas Sowell, 1993

Nothing is easier than spending public money. It does not appear to belong to anybody. The temptation is overwhelming to bestow it on somebody.

—Calvin Coolidge, 1927

The principle of spending money to be paid by posterity, under the name of funding, is but swindling futurity on a large scale.

—Thomas Jefferson, 1816

There is no such thing as a free lunch.

—Anonymous[40]

The state [...] can give nothing which it does not take from somebody.

—William Graham Sumner, 1883

The art of government consists of taking as much money as possible from one class of the citizens to give to the other.

—Voltaire, 1764

A government that robs Peter to pay Paul can always depend on the support of Paul.

—George Bernard Shaw, 1944[41]

One man's wage rise is another man's price increase.

—Harold Wilson, 1997

---

[38] As quoted, without a date, in several collections, starting not later than 1977.

[39] The so-called *Wicker's Law*—a natural extension of *Parkinson's Law*.

[40] Frequently misattributed to either Milton Freedman or Robert Heinlein; seems to first appear in press, in an unsigned newspaper piece, in 1938.

[41] Apparently rooted in a European proverb, "To strip Peter to clothe Paul".

The horse that draws best is most whipped.

—European proverb

Injurious is the gift that takes away freedom.

—Italian proverb

I fear the Danaans even when they bring gifts.[42]

—Virgil, ~ 20 BC

A government big enough to give you everything you want is a government big enough to take away everything you have.

—Gerald Ford, 1974

A claim for equality of material position can be met only by a government with totalitarian powers.

—Friedrich Hayek, 1976

Property must be secured, or liberty cannot exist.

—John Adams, 1790

Givers have to set limits because takers rarely do.

—Irma Kurtz, 2003

***

The government consists of a gang of men exactly like you and me. They have [...] no special talent for the business of government; they have only a talent for getting and holding office.

—H. L. Mencken, 1936

Let the people think they govern, and they will be governed.

—William Penn, 1682

If you can't convince 'em, confuse 'em.

—Harry S. Truman, 1948[43]

Every government is run by liars. Nothing they say should be believed.

—I. F. Stone, 1974

Believe nothing until it has been officially denied.

—Claud Cockburn, 1956

Since a politician never believes what he says, he is quite surprised to be taken at his word.

—Charles De Gaulle, 1962

---

[42] About the infamous *Trojan horse*, with the "Danaans" (Lat. *Danaos*) meaning the Greek invaders of the ancient Troy.

[43] Quoted in his address to farmers as "an old political trick".

It would be desirable if every Government, when it comes to power, should have its old speeches burnt.

—Philip Snowden, 1929

After eight years in Washington, I longed for the realism and sincerity of Hollywood.

—Fred Thompson, 2007[44]

The mystery of government is not how Washington works but how to make it stop.

—P. J. O'Rourke, 1991

Washington is a city of Southern efficiency and Northern charm.

—Anonymous[45]

Talk is cheap – except when Congress does it.

—Cullen Hightower[46]

No man's life, liberty or property are safe while the legislature is in session.

—Gideon J. Tucker, 1866

***

The more laws and orders are made prominent, the more thieves and robbers.

—Lǎozi, ~ 4th century BC

The more numerous the laws, the more corrupt the government.

—Tacitus, 117 AD

A state is much better governed when has only a few laws, which are strictly observed.

—René Descartes, 1628

Why keep on enacting laws when we already have more than we can break?

—Anonymous

***

We have the best politicians that money can buy.

—Edward Abbey, 1990

The louder he talked of his honor, the faster we counted our spoons.

—Ralph Waldo Emerson, 1860[47]

---

[44]He was an actor—and a US Senator.

[45]Frequently misattributed to John F. Kennedy, who only quoted it (in a 1961 speech) as a known joke.

[46]As quoted, without a date, in several reputable collections, starting not later than 1992.

[47]Apparently rooted in a much earlier (and much longer) remark by Samuel Johnson on a similar topic.

An honest politician is one who, when he is bought, will stay bought.

—*Anonymous*[48]

Politics is supposed to be the second oldest profession. I have come to realize that it bears a very close resemblance to the first.

—*Ronald Reagan, 1977*

***

Politics is perhaps the only profession for which no preparation is thought necessary.

—*Robert Louis Stevenson, 1882*

Politics is the art of the possible.

—*Otto Von Bismarck, 1867*

Politics is *not* the art of the possible. It consists in choosing between the disastrous and the unpalatable.

—*John Kenneth Galbraith, 1962*

Politics: a Trojan horse race.

—*Stanisław Jerzy Lec, 1957*

Politics is the skilled use of blunt objects.

—*Lester B. Pearson, 1982*

Politics is the art of looking for trouble, finding it whether it exists or not, diagnosing it incorrectly, and applying the wrong remedy.

—*Ernest Benn, 1930*

Practical politics consists in ignoring facts.

—*Henry Adams, 1907*

Politics is made up largely of irrelevancies.

—*Dalton Camp, 2001*

Politics is the art of preventing people from taking part in affairs which properly concern them.

—*Paul Valéry, 1943*

Politics and the fate of mankind are formed by men without ideals and without greatness. Those who have greatness within them do not go in for politics.

—*Albert Camus, 1962*

Political language [...] is designed to make lies sound truthful and murder respectable, and to give an appearance of solidity to pure wind.

—*George Orwell, 1950*

When a politician, on a subject implicating science says, 'the debate is over', you may be sure of two things: the debate is raging, and he is losing it.

---

[48]Frequently misattributed to American politician Simon Cameron (1799-1899).

—George Will, 2014

A good politician, under democracy, is quite as unthinkable as an honest burglar.

—H. L. Mencken, 1924

Ninety percent of the politicians give the other ten percent a bad reputation.

—Henry Kissinger, 1984

In politics you must always keep running with the pack. The moment that you falter and they sense that you are injured, the rest will turn on you like wolves.[49]

—R. A. Butler, 1989

Being in politics is like being a football coach. You have to be smart enough to understand the game, and dumb enough to think it's important.[50]

—Eugene McCarthy, 1967

Whoever could make two ears of corn or two blades of grass, to grow upon a spot of ground where only one grew before would deserve better of mankind and do more essential service to this country than the whole race of politicians put together.

—Jonathan Swift, 1726

The word 'politics' is derived from the word 'poly', meaning 'many', and the word 'ticks', meaning 'blood-sucking parasites'.

—Anonymous[51]

***

Too bad the only people who know how to run the country are busy driving cabs and cutting hair.

—George Burns, 1979

---

[49]I believe that, very unfortunately, this statement remains valid well beyond politics, covering even some scientists.

[50]Again, I believe the same may be said about many other human activities, including fashion, entertainment, and even some fields of science.

[51]This wonderful joke is widely attributed to some Larry Hardiman, but I could not confirm this authorship, and even identify this person.

On War and Peace

War is sweet to those who have no experience in it.

—*Pindar*, 446 BC

[War] is always an evil, never a good. We will not learn how to live together in peace by killing each other's children.

—*Jimmy Carter*, 2002

War is a kind of superstition; the pageantry of arms and badges corrupts the imagination of Man.

—*Percy Bysshe Shelley*, 1819

The higher animals engage in individual fights, but never in organized masses. Man is the only animal that deals with this atrocity of atrocities, War.

—*Mark Twain*, 1896

War is not an adventure. It is a disease. It is like typhus.

—*Antoine de Saint-Exupéry*, 1942

A hospital alone shows what war is, because its collection of bodies carries the true meaning of war.

—*Erich Maria Remarque*, 1929

War is cruelty, and you cannot refine it.

—*William Tecumseh Sherman*, 1864

When the sword is once drawn, the passions of men observe no bounds of moderation.

—*Alexander Hamilton*, 1788

Once lead this people into war and they will forget there ever was such a thing as tolerance.

—*Woodrow Wilson*, 1917

It is forbidden to kill; therefore all murderers are punished unless they kill in large numbers and to the sound of trumpets.

—*Voltaire*, 1771

K. K. Likharev (ed.), *Essential Quotes for Scientists and Engineers*,
https://doi.org/10.1007/978-3-030-63332-5_19

By whatever name men may call murder - murder always remains murder.

—Leo Tolstoy, 1909

Never think that war [...] is not a crime.

—Ernest Hemingway, 1946

***

War does not determine who is right - only who is left.

—Anonymous[1]

Truth is the first casualty of war.

—Mrs. Philip Snowden, 1915[2]

You need two dogs for a fight,
You need two wrongs for a war.

—Anonymous[3]

War settles nothing; [...] to win a war is as disastrous as to lose one.

—Agatha Christie, 1977

We hear war called murder. It is not: it is suicide.

—Ramsey MacDonald, 1930

You can no more win a war than you can win an earthquake.

—Jeannette Rankin, 1974

The statesman who yields to war fever must realize that once the signal is given, he is no longer the master of policy but the slave of unforeseeable and uncontrollable events.

—Winston Churchill, 1930

***

War is an instrument entirely inefficient toward redressing wrong; and multiplies, instead of indemnifying losses.

—Thomas Jefferson, 1798

The direct use of physical force is so poor a solution to the problem of limited resources that it is commonly employed only by small children and great nations.

—David Friedman, 1973

Violence is the last refuge of the incompetent.

—Isaac Asimov, 1951

---

[1]Frequently misattributed to Bertrand Russell.

[2]Samuel Johnson expressed a similar idea, but in a much longer form, as early as in 1758. Variants of this maxim are frequently misattributed to Aeschylus (525-456 BC), and Hiram Johnson (1866-1945). It was paraphrased in 1928 by Arthur Ponsonby, without a reference.

[3]Perhaps rooted in a Dutch proverb, "When two quarrel, both are in the wrong".

War is only a cowardly escape from the problems of peace.

—Thomas Mann, 1954

I cease not to advocate peace. It may be on unjust terms, but even so it is more expedient than the justest of civil wars.

—Cicero, ~44 BC

The most disadvantaged peace is better that the most just war.

—Erasmus, 1508

***

[Many] go to war because they don't want to be a hero.

—Tom Stoppard, 1972

Unhappy is the land that needs heroes.

—Bertold Brecht, 1939

As a rule, heroism is due to a lack of reflection, and thus it is necessary to maintain a mass of imbeciles. If they once understand themselves the ruling men will be lost.

—Ernest Renan, 1878

If my soldiers were to begin to think, not one would remain in the ranks.

—Frederick the Great[4]

War will never cease until babies begin to come into the world with larger cerebrums and smaller adrenal glands.

—H. L. Mencken, 1956

I'm glad I didn't have to fight in any war.
I'm glad I didn't have to pick up a gun.
I'm glad I didn't get killed or kill somebody.
I hope my kids enjoy the same lack of manhood.

—Tom Hanks, 1998

***

You can't say that civilization don't advance, however, for in every war they kill you in a new way.

—Will Rogers, 1929

Who desires peace, should prepare for war.

—Vegetius, fourth century AD[5]

---

[4]As quoted, without a date, by J. A. Houlding in 1981.

[5]This infamous advice (sometimes seconded by scientists—see the next quote) was followed by governments of all countries for almost two millennia. Now we know the disastrous results of the resulting arms race all too well.

If we stay strong, then I believe we can stabilize the world and have peace based on force.

—*Edward Teller*, 1958

The greatest evil that can oppress civilized people derives from [...] the never-ending arming for future war.

—*Immanuel Kant*, 1795

Great armaments lead inevitably to war.

—*Edward Grey*, 1925

Why does the Air Force need expensive new bombers? Have the people we've been bombing over the years been complaining?

—*George Wallace*[6]

Why I oppose the nuclear-arms race: I prefer the human race.

—*Edward Abbey*, 1990

War is much too serious a matter to be entrusted to the military.

—*Georges Clemenceau*, 1919[7]

***

Wars teach us not to love our enemies, but to hate our allies.

—*W. L. George*[8]

The enemy is anybody who's going to get you killed, no matter which side he's on.

—*Joseph Heller*, 1961

All men are brothers. [...] All wars are civil wars.

—*François Fénelon*, 1699

Our true nationality is mankind.

—*H. G. Wells*, 1920

I am first a man and only then a Frenchman.

—*Montesquieu*, 1721

Love for your fatherland is a wonderful thing, but there is something even more beautiful – this is love for the truth.

—*Pyotr Chaadaev*, 1837

Love for one's country which is not part of one's love for humanity is not love, but idolatrous worship.

—*Erich Fromm*, 1955

---

[6]As quoted, without a date, by Hakeem Shittu and Callie Query in 2006.

[7]This maxim, attributed by some to Charles de Talleyrand and by others to Aristide Briand, was later repeated by Charles De Gaulle about politics and politicians, by Iain Macleod about history and historians, and by John Archibald Wheeler about philosophy and philosophers. (I believe that in sciences, similarly, theory is much too serious a matter to be entrusted to dedicated theorists).

[8]As quoted, without a date, in several reputable sources, starting not later than 1970.

Patriotism is the last refuge of a scoundrel.

—Samuel Johnson, 1775

Patriotism is, fundamentally, a conviction that a particular country is the best in the world because you were born in it.

—George Bernard Shaw, 1893

Patriotism, as I see it, is often an arbitrary veneration of real estate above principles.

—George Jean Nathan, 1931

Patriotism is an unnatural, irrational, and harmful feeling.

—Leo Tolstoy, 1900

Patriotism varies, from a noble devotion to a moral lunacy.

—Dean Inge, 1919

Patriots always talk of dying for their country but never of killing for their country.

—Bertrand Russell, 1962

There is only one faith for which large masses of us are prepared to die and kill, and this faith is nationalism.

—Aldous Huxley, 1958

Born in iniquity and conceived in sin, the spirit of nationalism has never ceased to bend human institutions to the service of dissention and distress.

—Thorstein Veblen, 1923

Nationalism is an infantile disease. It is the measles of mankind.

—Albert Einstein, 1918

Nationalism is power hunger tempered by self-deception.

—George Orwell, 1945

Nationalism is our form of incest, is our idolatry, is our insanity.

—Erich Fromm, 1955

Nationalism [...] is like cheap alcohol. First, it makes you drunk, then it makes you blind, and then it kills you.

—Dan Fried, 2007

***

I confess that I cannot understand how we can plot, lie, cheat and commit murder abroad and remain humane, honorable, trustworthy and trusted at home.

—Archibald Cox[9]

---

[9] As quoted, without a date, in several generally reliable sources, starting not later than 1977, i.e. during the author's lifetime.

I have ever deemed it fundamental for the United States, never to take active part in the quarrels of Europe.

—Thomas Jefferson, 1823

I have always given it as my decided opinion, that no nation has a right to intermeddle into the internal concerns of another; that every one had a right to form and adopt whatever government they liked best to live under themselves.

—George Washington, 1796

Every nation gets the government it deserves.

—Joseph de Maistre, 1811

No nation is fit to sit in judgment upon any other nation.

—Woodrow Wilson, 1915

A nation is a historic group of men of recognizable cohesion, held together by a common enemy.

—Theodore Herzl, 1896

A nation is a society united by delusions about its ancestry and by common hatred of its neighbors.

—Dean Inge, 1948

Every nation ridicules other nations, and all are right.

—Arthur Schopenhauer, 1851

Every nation is selfish, and every nation considers its selfishness sacred.

—Antoine de Saint-Exupéry, 1944

The great nations have always acted like gangsters, and the small nations like prostitutes.

—Stanley Kubrick, 1963

***

It is easier to make war than to make peace.

—Georges Clemenceau, 1919[10]

Nobody can have peace longer than his neighbor pleases.

—Danish proverb

Peace, *n.* In international affairs, a period of cheating between two periods of fighting.

—Ambrose Bierce, 1911

If any question why we died,
Tell them, because our fathers lied.

—Rudyard Kipling, 1918

---

[10]I believe this is a very poor excuse for the disastrous Versailles Peace he had authored and enforced—almost single-handedly.

All governments lie, but disaster lies in wait for countries whose officials smoke the same hashish they give out.

—I. F. Stone, 1967

How is the world ruled and led to war? Diplomats tell lies to journalists and then believe these lies when they see them in print.

—Karl Kraus, 1976

Lying increases the creative faculties, expands the ego, lessens the frictions of social contacts, and cultivates memory.

—Clare Booth Luce, 1970[11]

Diplomacy, *n*. The patriotic art of lying for one's country.

—Ambrose Bierce, 1906

Diplomacy: lying in state.

—Oliver Herford, 1906[12]

Diplomacy without arms is like music without instruments.

—Frederick the Great[13]

Diplomacy frequently consists in soothingly saying 'Nice doggie' until you have a chance to pick up a rock.

—Walter S. Trumbull, 1925[14]

***

The only sane policy for the world is that of abolishing war.

—Linus Pauling, 1962

If everyone demanded peace instead of another television set, then there'd be peace.

—John Lennon, 1984

Let's not have any more wars to end all war.[15]

—William Feather, 2008

Hobbes: 'How come we play war and not peace?'
Calvin: 'Too few role models.'

—Bill Watterson, 1987

---

[11]She was a politician *and* a diplomat, so we should believe her (in this :-).

[12]Note that in many citations, the author's last name is misspelled as "Hereford".

[13]As quoted, without a date, in several authoritative history books.

[14]Sometimes misattributed to others, most frequently to Will Rogers.

[15]This is in reference to an initially popular name of WWI, *War That Will End War*, which had been in particular used by H. G. Wells for the title of his 1914 book—for that he later (in 1934) apologized.

If man does find the solution for world peace it will be the most revolutionary reversal of his record we have ever known.

—George Marshall, 1945

I can picture in my mind a world without war, a world without hate. And I can picture us attacking that world, because they'd never expect it.

—Jack Handey, 1992

On Success and Glory, Good and Bad

We must believe in luck. For how else can we explain the success of those we don't like?

—Jean Cocteau, 1955

It is not enough to succeed. Others must fail.

—Gore Vidal, 1976

Modest successes are better known as failures.

—Frederic Raphael, 2001

Success usually comes to those who are too busy to be looking for it.

—Anonymous[1]

Success is to be measured not so much by the position that one has reached in life as by the obstacles which he has overcome while trying to succeed.

—Booker T. Washington, 1901

There is only one success [...] - to be able to spend your life in your own way.

—Christopher Morley, 1922

***

A celebrity is a person who works hard all his life to become well known, then wears dark glasses to avoid being recognized.

—Fred Allen, 1954

The celebrity is a person who is known for his well-knownness.

—Daniel J. Boorstin, 1961

People seldom become famous for what they say, until after they are famous for what they've done.

—Cullen Hightower[2]

---

[1]Frequently misattributed to Henry David Thoreau.

[2]As quoted, without a date, in several generally reliable sources, including *Wikipedia*.

K. K. Likharev (ed.), *Essential Quotes for Scientists and Engineers*,
https://doi.org/10.1007/978-3-030-63332-5_20

Vainglory blossoms, and bears no fruit.

—Spanish proverb

Live unknown.

—Epicurus[3]

I'm Nobody! [...]
How dreary – to be – Somebody!

—Emily Dickinson, ~ 1861

Less glory is more liberty.

—Albert Pike, 1871

It is graceless to be famous.

—Boris Pasternak, 1956

Some people make headlines while others make history.[4]

—Philip Elmer-DeWitt[5]

***

The ends justify the means.

—Ovid, ~ 20 BC

The good ends do not make the means good.

—Gervase Babington, 1583

The ends cannot justify the means, for the simple and obvious reason that the means employed determine the nature of the ends produced.

—Aldous Huxley, 1937

If you have no aim, you need not worry about the means.

—Fausto Cercignani, 2004

Where there is no shame, there is no honor.

—European proverb

He alone is great and happy who requires neither to command nor to obey in order to secure his being of some importance in the world.

—Johann Wolfgang von Goethe, 1773

The greatest power available to man is not to use it.

—Meister Eckhart[6]

---

[3]Apparently, this statement has become first known from its criticism by Plutarch in the first century AD.

[4]On the same general theme, there is a much more prosaic Spanish proverb: "Some have the fame, and others card the wool".

[5]As quoted, without a date, in several generally reliable sources, starting not later than 1998.

[6]As quoted, without a date, in several comprehensive collections of his sermons.

There is no real greatness without simplicity, goodness, and truth.

—Leo Tolstoy, 1869

Live among men as if God saw you; speak to God as if men heard you.

—Seneca, ~65 AD[7]

A good Life is the only Religion.

—Thomas Fuller, 1732

What you do not want done to yourself, do not do to others.

—Confucius, fifth century BC[8]

Morality is the observance of the rights of others.

—Dagobert D. Runes, 1966

Aim above morality. Be not simply good; be good for something.

—Henry David Thoreau, 1848

Every man is worth just so much as the things are worth about which he busies himself.

—Marcus Aurelius, 1st century AD

***

The greatest minds are capable of the greatest vices as well as of the greatest virtues.

—René Descartes, 1637

It has been my experience that folks who have no vices have very few virtues.

—Abraham Lincoln, 1866

The problem with people who have no vices is that generally you can be pretty sure they're going to have some pretty annoying virtues.

—Elizabeth Taylor, 2000

Virtue is indeed its own noblest reward.

—Silius Italicus, first century AD[9]

---

[7] A much later rephrase by Cardinal Francis Spellman is also frequently cited.

[8] Probably not original even in that time, this "Golden Rule" of morality was reproduced by Socrates and Isocrates during the next century, in the Indian *Mahabharata* a century after that, and by innumerable prophets, and preachers later on. I am not aware of any essential addition to this wisdom, made by any religion. For example, Hillel the Elder, one of the most prominent Jewish religious scholars, said, "That which is hateful to you, do not do to your fellow. That is the whole Law; the rest is the explanation." Note that the altruistic behavior implied by the rule may be explained as a result of the biological evolution—see the last quotes in the section *On Faith and Religion*.

[9] Possibly rooted in Plato's "Virtue was sufficient of herself for happiness"; sometimes misattributed to Izaak Walton, Matthew Prior, Henry More, John Gay and many later authors, frequently shortened to "Virtue is its own reward".

Virtue is its own punishment.

—Anonymous[10]

No good deed goes unpunished.

—Clare Booth Luce, 1956[11]

There is no moral precept that does not have something inconvenient about it.

—Denis Diderot[12]

What is light on your conscience, is often heavy on your back.

—Danish proverb

Do what you feel in your heart to be right - for you 'll be criticized anyway. You'll be 'damned if you do, and damned if you don't'.[13]

—Eleanor Roosevelt, 1944

Always do right. This will gratify some people and astonish the rest.

—Mark Twain, 1901

Nothing astonishes men so much as common sense and plain dealing.

—Ralph Waldo Emerson, 1841

***

Conscience is God's presence in man.

—Emanuel Swedenborg, 1756

Conscience is, in most men, an anticipation of the opinion of others.

—Henry Taylor, 1836

Conscience is the inner voice that warns us somebody may be looking.

—H. L. Mencken, 1949

It is far more impressive when others discover your good qualities without your help.

—Judith Martin, 1991

The greatest pleasure I know, is to do a good action by stealth, and to have it found out by accident.

—Charles Lamb, 1834

***

You become responsible forever for what you have tamed.

—Antoine de Saint-Exupéry, 1943

---

[10]Sometimes attributed to Aneurin Bevan, but I could not find a reliable confirmation of his authorship.

[11]Sometimes attributed to Oscar Wilde and Billy Wilder, without written evidence.

[12]As quoted, without a date, in several reliable sources, starting not later than 1998.

[13]The expression in quote signs has been traced back to sermons by Lorenzo Dow (in the 1830s).

When you helped somebody, right away you were responsible for that person. And things always followed for which you were never prepared.

—Martha Brooks, 2003

***

You are responsible for the talent that has been entrusted to you.

—Henri-Frédéric Amiel, 1883

We can't take any credit for our talents. It's how we use them that counts.

—Madeleine L'Engle, 1962

We work in the dark – we do what we can – we give what we have.

—Henry James, 1893

Give what you have. To someone, it may be better than you dare think.

—Henry Wadsworth Longfellow, 1849

A hero is a man who does what he can. The others do not do it.

—Romain Rolland, 1904

Each river does its best for the sea.

—Czech proverb

We must cultivate our garden.

—Voltaire, 1759

In doing what we ought we deserve no praise, because it is our duty.

—Saint Augustine[14]

When we have done our best, we should wait the result in peace.

—John Lubbock, 1887

Do your duty and leave the rest to the gods.

—Pierre Corneille, 1639

Duty is ours, events are God's.

—Matthew Henry, 1710[15]

Our duty is to crow, and then the dawn may or may not come.

—Vladimir Bogomolov, 1973

---

[14]As quoted, without a date, in several reliable sources, starting not later than 1824.

[15]A much later paraphrase of this maxim (with "results" instead of "events") by John Quincy Adams is quoted more often than the original.

On the World and Time, Life and Death

We cross our bridges when we come to them and burn them behind us, with nothing to show for our progress except a memory of the smell of smoke, and a presumption that once our eyes watered.

—Tom Stoppard, 1966

All changes, even the most longed for, have their melancholy; for what we leave behind us is a part of ourselves; we must die to one life before we can enter another.

—Anatole France, 1881

Nothing is as far away as one minute ago.

—Anonymous[1]

***

Nothing exists except atoms and empty space; everything else is opinion.

—Democritus, third century BC

God made everything out of nothing, but the nothingness shows through.

—Paul Valéry, 1941

Eternal nothingness is OK if you happen to be dressed for it.

—Woody Allen, 1971

***

The whole world is an enigma, a harmless enigma that is made terrible by our own mad attempt to interpret it as though it had an underlying truth.

—Umberto Eco, 2014

The most incomprehensible thing about the world is that it is comprehensible.

—Albert Einstein, 1955

---

[1]As quoted, without a date, in many collections, with attributions either to Jef Mallett or to Jim Bishop. Unfortunately, I was unable to resolve this dilemma.

© The Author(s), under exclusive license to Springer Nature Switzerland AG 2021   197
K. K. Likharev (ed.), *Essential Quotes for Scientists and Engineers*,
https://doi.org/10.1007/978-3-030-63332-5_21

Nature, in her indifference, makes no distinction between good and evil.

—Anatole France, 1914

Nature is not cruel, only pitilessly indifferent.

—Richard Dawkins, 1995

This universe is not hostile, nor yet is it friendly. It is simply indifferent.

—John Haynes Holmes, 1932.

***

All the world's a stage,
And all the men and women merely players.

—William Shakespeare, 1599

All the world's a stage,
And most of us are desperately under-rehearsed.

—Seán O'Casey, 1964[2]

All the world's a cage,
And all the men and women are merely hamsters.[3]

—Tony Vigorito, 2017

It is impossible to imagine the universe run by a wise, just and
omnipotent God, but it is quite easy to imagine it run by a board of gods. If such a board
actually exists it operates exactly like the board of a corporation that is losing money.
—H. L. Mencken, 1956

***

When I was born I wept, and every day brings a reason why.

—Spanish proverb

Maybe this world is another planet's hell.

—Aldous Huxley, 1954

Who made the world I cannot tell;
'Tis made, and here am I in hell.

—A. E. Housman, 1936

The world's a sorry wench, akin
To all that's fail and frightful:
The world's as ugly, ay, as Sin –
And almost as delightful.

—Frederick Locker-Lampson, 1862

---

[2]I am not fully confident this publication was the first one.

[3]This is an extension of the earlier first line by Jeanne Phillips.

The world's a fine place for those who go out to take it; there's lots of unknown stuff in it yet.

—John Galsworthy, 1901

The effort to understand the universe is one of the very few things that lifts human life a little above the level of farce, and gives it some of the grace of tragedy.

—Steven Weinberg, 1993

The actual tragedies of life bear no relation to one's preconceived ideas. In the event, one is always bewildered by their simplicity, their grandeur of design, and by that element of the bizarre which seems inherent in them.

—Jean Cocteau, 1955

Tragedy is the difference between what is and what could have been.

—Abba Eban, 1963

Man is the only animal that laughs and weeps, for he is the only animal that is struck with the difference between what things are and what they ought to be.

—William Hazlitt, 1819

The mark of your ignorance is the depth of your belief in injustice and tragedy. What the caterpillar calls the end of the world, the master calls a butterfly.

—Richard Bach, 1977

This world is a comedy to those that think; a tragedy to those that feel.

—Horace Walpole, 1769[4]

Life is a festival only to the wise.

—Ralph Waldo Emerson, 1841

Life is too tragic for sadness: Let us rejoice.

—Edward Abbey, 1990

Life would be tragic if it weren't funny.

—Stephen Hawking, 2004

Life does not cease to be funny when people die any more than it ceases to be serious when people laugh.

—George Bernard Shaw, 1911

Life is far too important a thing ever to talk seriously about it.

—Oscar Wilde, 1882

Not a shred of evidence exists in favor of the idea that life is serious.

—Brendan Gill, 1975

The more you find out about the world, the more opportunities there are to laugh at it.

—Bill Nye, 1892[5]

---

[4]Sometimes misattributed to Jean de la Bruyère.

[5]Frequently misattributed to "Bill Nye the Science Guy" (William Sanford Nye, 1955–present).

In the end, everything is a gag.

—Charlie Chaplin, 1974

***

What is life but a series of inspired follies?

—George Bernard Shaw, 1912

Life is just one damn thing after another.

—Elbert Hubbard, 1909

Life is a four-letter word.

—Lenny Bruce[6]

Life is never fair.

—Oscar Wilde, 1895

Life being what it is, one dreams of revenge.

—Paul Gauguin, 1903

Life *is* pain, Highness! Anyone who says differently is selling something[7]

—William Goldman, 1987

Life is one long struggle in the dark.

—Lucretius, first century BC

Life is a warfare.

—Seneca, 64 AD

Life is a struggle, but not a warfare, it is a day's labor, […] where we may think and sing and enjoy as we work.

—John Burroughs, 1913

Life is an offensive directed against the repetitious mechanism of the Universe.

—Alfred North Whitehead, 1933

Life is like an onion. You peel it off one layer at a time, and sometimes you weep.

—Carl Sandburg[8]

Life is made up of sobs, sniffles and smiles, with sniffles predominating.

—O. Henry, 1906

Life is a protracted struggle against the Adversary, who is man himself.

—Max Lerner, 1959

---

[6] As quoted, without a date, in several reputable sources including *Wikiquote*.

[7] Cf. "Life's good", the LG corporation's motto.

[8] As quoted, without a date, in several reputable collections, starting not later than 1970.

Life is not easy for any of us. But what of that?

—Marie Curie, 1937

Life is easier to take that you'd think; all that is necessary is to accept the impossible, do without indispensable, and bear the intolerable.

—Kathleen Norris, 1928

***

Life is a hospital where each patient is possessed by the desire to change his bed.

—Charles Baudelaire, 1862

Life is like a department store: it has everything but what you are looking for.

—Emil Krotky, 1966

Life is a zoo in a jungle.

—Peter De Vries, 1967

Life is a sexually transmitted disease.

—Anonymous[9]

Life is not an exact science, it is an art.

—Samuel Butler, 1912

Life is something that everyone should try at least once.

—Henry J. Tillman, 2009

In spite of the cost of living, it's still popular.

—Laurence J. Peter, 1977[10]

Life is a long lesson in humility.

—J. M. Barrie, 1891

***

This life is a test – it is only a test. If it had been an actual life, you would have received further instructions on where to go and what to do.

—Jack Kornfield, 1993[11]

Life can only be understood backwards. But [...] it must be lived forwards.

—Søren Kierkegaard, 1844

Life is fired at us point-blank.

—José Ortega y Gasset, 1957

---

[9]To this joke, popular at least since 1980, R. D. Laing added (in 1985): "...with the 100% mortality rate".

[10]Sometimes misattributed to Kathleen Norris.

[11]I have seen claims that this joke had been used by others earlier than 1993, but was unable to confirm them.

Life is half spent before one knows what life is.

—French proverb

Life is like playing a violin solo in public and learning the instrument as one goes on.

—Samuel Butler, 1895

Life is a maze in which we take the wrong turning before we have learnt to walk.

—Cyril Connolly, 1933

Life is the only art that we are required to practice without preparation.

—Lewis Mumford, 1951

Life is the art of drawing without an eraser.

—John W. Gardner, 1992[12]

***

Life is something that happens when you can't get to sleep.

—Fran Lebowitz, 1978

Life is what happens to us while we are making other plans.

—Allen Saunders, 1957[13]

Life flows on within you and without you.

—George Harrison, 1967

***

Life is short, art long, opportunity fleeting, experimentation perilous, and judgment difficult.[14]

—Hippocrates[15]

Life is full of misery, loneliness, and suffering - and it's all over much too soon.

—Woody Allen, 1975

Life isn't long enough for love and art.

—W. Somerset Maugham, 1919

Life is too short to stuff a mushroom.

—Shirley Conran, 1975

Growing old is no more than a bad habit which a busy man has no time to form.

—André Maurois, 1939

---

[12]Actually, he began this line as "Somebody said that ...".

[13]Frequently misattributed to others, including John Lennon, who included this line into his song (recorded in 1980). A more recent popular (anonymous?) version of this maxim is: "Life is what passes by while you're texting.".

[14]From the context, the word "art" here means the art and science of medicine.

[15]This sentence was popular already among the Romans, in a shorter Latin form: "*Ars longa, vita brevis*".

The secret of staying young is to live honestly, eat slowly, and lie about your age.

—Lucille Ball, 1983

Live forever or die in the attempt.

—Joseph Heller, 1961

The proper function of man is to live, not to exist. I shall not waste my days in trying to prolong them.

—Jack London, 1916

Quit worrying about your health. It'll go away.

—Robert Orben, 1991

Old age isn't so bad when you consider the alternative.

—Maurice Chevalier, 1960

I detest life-insurance agents: they always argue that I shall some time die, which is not so.

—Stephen Leacock, 1910

The majority of people, though they do not know what to do with this life, long for another that shall have no end.

—Anatole France, 1914

Millions long for immortality who don't know what to do with themselves on a rainy Sunday afternoon.

—Susan Ertz, 1943

***

A sage dies in time.

—Maxim Gorky, 1908

Tired of all this, for restful death I cry.

—William Shakespeare, 1609

Death [is] the only thing we haven't succeeded in completely vulgarizing.

—Aldous Huxley, 1936

On the plus side, death is one of the few things that can be done just as easily lying down.

—Woody Allen, 1976

When I die, I want to go peacefully in my sleep like my grandfather. Not screaming in terror, like the passengers in his car.

—Jack Handey, 1993

I'm having a glorious old age. One of my greatest delights is that I have outlived most of my opposition.

—Maggie Kuhn, 1978

[He] is one of those people who would be enormously improved by death.

—Saki, 1914

I have never killed any one, but I have read some obituary notices with great satisfaction.

—*Clarence Darrow*, 1932

I didn't attend his funeral, but sent a nice letter saying that I approved of it.

—*Anonymous*[16]

They say such nice things about people at their funerals that it makes me sad to realize that I'm going to miss mine by just a few days.

—*Garrison Keillor*, 2005

It is not death that a man should fear, but he should fear never beginning to live.

—*Marcus Aurelius*, second century AD

A man should be mourned at his birth, not his death.

—*Montesquieu*, 1721

The tragedy of modern man is not that he knows less and less about the meaning of his own life, but that this bothers him less and less.

—*Václav Havel*, 1988

What remains of most people, is only a dash between two dates.

—*Georgi Polonsky,* 1968[17]

Let life happen to you.

—*Rainer Maria Rilke*, 1904

You can't do anything about the length of your life, but you can do something about its width and depth.

—*Evan Esar*, 1924

Life has a value only when it has something valuable as its object.

—*Georg Wilhelm Freidrich Hegel*, 1832

It's not the things we do in life that we regret on our death bed, it is the things we do not.

—*Randy Pausch*, 2008[18]

It matters not how a man dies, but how he lives.

—*Samuel Johnson*, 1769

You only live once, but if you do it right, once is enough.

—*Anonymous*[19]

---

[16]This popular joke, frequently misattributed to Mark Twain, is actually just a paraphrase of a longer statement by Ebenezer R. Hoar, made in 1884.

[17]Polonsky was the screenwriter of Stanislav Rostovsky's movie *We'll Live Till Monday*, in that this phrase is pronounced, but I am not confident that it was fully original at that time. Later (in 1996) this idea was used by Linda Ellis as the basis of her popular poem *The Dash*.

[18]From the author's famous *Last Lecture*, given knowing of his impending death of cancer in a few months.

[19]I have seen a rather questionable attribution of this quote to Mae West.

We should remember our dying, and try so to live that our death brings no pleasure to the world.

—John Steinbeck, 1952

Remembering that I'll be dead soon is the most important tool I've ever encountered to help me make the big choices in life.

—Steve Jobs, 2005

I know of no more encouraging fact than the unquestionable ability of man to elevate his life by a conscious endeavor.

—Henry David Thoreau, 1854

The great end of life is not knowledge but action.

—Thomas Henry Huxley, 1877

It is absolutely a crime for any man to die possessed of useful knowledge in which nobody else shares.

—J. A. L. Waddell et al., 1933

What we have done for ourselves alone dies with us; what we have done for others and the world remains, and is immortal.

—Albert Pike, 1860

Only a life lived for others is a life worthwhile.

—Albert Einstein, 1932

***

Lost wealth may be replaced by industry, lost knowledge by study, lost health by temperance or medicine, but lost time is gone for ever.

—Samuel Smiles, 1864

A man who dares to waste one hour of time has not discovered the value of life.

—Charles Darwin, 1836

We have only this moment, sparkling like a star in our hand and melting like a snowflake. Let us use it before it is too late.

—Marie Beynon Ray, 1953[20]

We all have such a finite time to leave the world better than we found it.

—Dave Kellett, 2012

Time is the most valuable thing a man can spend.[21]

—Theophrastus[22]

---

[20]In several quote collections, the second name of the author is misspelled as "Beyon".

[21]Here I have to mention the extremely popular European proverb "Time is money" (frequently misattributed to Benjamin Franklin), but only to register my humble opinion that it is rather shallow.

[22]First quoted by Diogenes Laërtius in third century AD.

Lost time is never found again.

—Benjamin Franklin, 1748

Time is a trust, and for every minute of it you will have to account.

—John Lubbock, 1895

If you can fill the unforgiving minute
With sixty seconds' worth of distance run,
Yours is the Earth and everything that's in it,
And – which is more - you'll be a Man, my son!

—Rudyard Kipling, 1896

One should count each day a separate life.

—Seneca, ∼65 AD

Thing of today, accomplished today.

—Chinese proverb

Do not wait for the last judgment. It takes place every day.

—Albert Camus, 1957

Live neither in the past nor in the future, but let each day's work absorb your entire energies, and satisfy your widest ambition.

—William Osler, 1899

Happy is the man who can recognize in the work of to-day a connected portion of the work of life and an embodiment of the work of eternity.

—James Clerk Maxwell, 1854

Life itself is brief, and that is what charges each day with such ridiculous beauty.

—Garrison Keillor, 1990

If a man has a great deal to put into them, a day will have a hundred pockets.

—Friedrich Nietzsche, 1888

I still find each day too short for all the thoughts I want to think, all the walks I want to take, all the books I want to read, and all the friends I want to see.

—John Burroughs, 1913

I arise in the morning torn by a desire to improve (or save) the world, and a desire to enjoy (or savor) the world. This makes it hard to plan the day.

—E. B. White, 1969

The intellect of man is forced to choose
Perfection of the life, or of the work.

—William Butler Yeats, 1933

***

No man loves life like him that's growing old.[23]

—*Sophocles*, fifth century BC

Like as the waves make towards the pebbl'd shore,
So do our minutes, hasten to their end.

—*William Shakespeare*, 1609

***

If thou wouldest win Immortality of Name, either do things worth the writing, or write things worth the reading.

—*Thomas Fuller*, 1727[24]

Seek not, my soul, immortal life, but make the most of the resources that are within your reach.

—*Pindar*, 498 BC

Reason is immortal.

—*Pythagoras*[25]

Immortality is to labor at an eternal task.

—*Ernest Renan*, 1890

If you want immortality, make it!

—*Joaquin Miller*, 1908

Since man is mortal, the only immortality possible for him is to leave something behind him that is immortal.

—*William Faulkner*, 1958

If all else fails, immortality can always be assured by spectacular error.

—*John Kenneth Galbraith*, 1975

***

[French chemists Auguste Laurent and Charles Frédéric Gerhardt] died, ignored by most; they never sought nor found public favor, for high roads never lead there.

—*Le Moniteur Scientifique Du Docteur Quesneville* (editorial), 1862

Those only deserve a monument who do not need one; that is, who have raised themselves a monument in the minds and memories of men.

—*William Hazlitt*, 1823

---

[23]A much later similar confession by Frank Lloyd Wright is also popular.

[24]Frequently misattributed to Benjamin Franklin, who just paraphrased it (in 1738) in his *Poor Richard's Almanac*.

[25]First quoted by Diogenes Laërtius in the third century AD.

I have built a monument more lasting than bronze
and set higher than the pyramids of the kings.
It cannot be destroyed by gnawing rain
or wild north wind, by the procession
of immeasurable years or by the flight of time

—*Horace*, 23 BC[26]

The monuments of wit survive the monuments of power.

—*Francis Bacon*, 1595

If you seek Hamilton's monument [in Washington, DC], look around. You are living in it.
We […] live in Hamilton's country.

—*George Will*, 1992

I would rather have men asking why I have no statue than why I have one.

—*Cato*[27]

[Lise Meitner's] work was crowned by the Nobel Prize for Otto Hahn.

—*Renate Feyl*, 1981

The seeds we have sown are germinating under others' names.

—*Alexander Gorodnitsky*, 1959[28]

I must consider the organizer as more important that the discoverer.

—*Wolfgang Ostwald*, 1927

In science the credit goes to the man who convinces the world, not to the man to whom the
idea first occurs. Not the man who finds a grain of new and precious quality but to him who
sows it, reaps it, grinds it and feeds the world on it.

—*Francis Darwin*, 1914

Discoveries are usually not made by one man alone, but […] many brains and many hands
are needed before a discovery is made for which one man receives the credit.

—*Henry E. Sigerist*, 1951

It is amazing what you can accomplish if you do not care who gets the credit.

—*Anonymous*[29]

---

[26]From his *Ode III.30*; translation by David West.

[27]As quoted, without a date, by Plutarch in the early second century AD.

[28]A verse from a popular Russian song, in a fair but prosaic translation.

[29]This wisdom and its variants are attributed to many, including Ronald Reagan and Harry S.
Truman, with the earliest (circa 1863) attribution to some Farther Stirckland, but I could not find a
reliable confirmation of any of these authorships.

Conclusion: Farewell

A book of quotations can never be complete.

—Anonymous[1]

A poem is never finished, it is only abandoned.

—W. H. Auden, 1967[2]

Fare thee well! and if forever,
Still forever, fare thee well.

—Lord Byron, 1816

---

[1]I have seen this quote attributed to some Robert M. Hamilton, but could not find a reliable confirmation of this authorship.

[2]This is a succinct paraphrase of an earlier and much longer sentence by Paul Valéry. (Auden cited him as the author.) It was later repeated, about books, by Gene Fowler.

K. K. Likharev (ed.), *Essential Quotes for Scientists and Engineers*,
https://doi.org/10.1007/978-3-030-63332-5_22

Bibliography

Major Quotation Sources (on Top of Original Publications)

A. Books

T. Augarde, *The Oxford Dictionary of Modern Quotation* (Oxford U. Press, 1991)

J. Bartlett, *Familiar Quotations*, 16th ed. (Little, Brown and Company, 1992)

H.G. Bohn, *A Polyglot of Foreign Proverbs* (Franklin Classics, 2018)

E.E. Brussell, *Webster's New World Dictionary of Quotable Definitions*, 2nd ed. (Prentice-Hall, 1988)

W.F. Bynum, R. Potter, *Oxford Dictionary of Scientific Quotations* (Oxford U. Press, 2005)

B. Evans, *The Dictionary of Quotations* (Random House, 1993)

C.C. Gaither, A.E. Cavazos-Gaither, *Practically Speaking* (IOP, 1999)

C.C. Gaither, A.E. Cavazos-Gaither (eds.), *Gaither's Dictionary of Scientific Quotations*, in 2 vols. (Springer, 2007)

R. Keyes, *The Quote Verifier* (Griffin, 2006)

E. Knowles, *Oxford Dictionary of Modern Quotations*, 3rd ed. (Oxford U. Press, 2007)

E. Knowles, *Oxford Dictionary of Quotations*, 8th ed. (Oxford U. Press, 2014)

A.L. Mackay, *A Dictionary of Scientific Quotations*, 2nd ed. (IOP, 1991)

E.C. McKenzie, *14,000 Quips & Quotes* (Wings Books, 1984)

L.J. Peter, *Peter's Quotations* (William Morrow, 1977)

G. Seldes, *The Great Thoughts* (Ballantine Books, 1985)

J. Speake (ed.) *The Oxford Dictionary of Proverbs*, 6th ed. (Oxford U. Press, 2015)

E. Strauss, *Dictionary of European Proverbs* (Routledge, 1994)

B. Swainson (ed.) *The Encarta Book of Quotations* (Bloomsbury/St. Martins, 2000)

The Oxford Dictionary of Quotations, 3rd ed. (Oxford U. Press, 1979)

A. Watson (ill.) *Russian Proverbs* (Literary Licensing, 2011)

B. Web Sites

Goodreads, https://goodreads.com/quotes/

Libquotes, https://libquotes.com/

Quote Investigator, https://quoteinvestigator.com/

© The Editor(s) (if applicable) and The Author(s), under exclusive license to Springer Nature Switzerland AG 2021
K. K. Likharev (ed.), *Essential Quotes for Scientists and Engineers*,
https://doi.org/10.1007/978-3-030-63332-5

The Quotations Page, https://www.quotationspage.com/
Wikiquote, https://en.wikiquote.org/
Wisdom Quotes, https://wisdomquotes.com/

Author Index

The index does not include the authors who merely quoted the original work, and the persons to whom the quotes were misattributed.

Collective authorships

[1]French scientific journal published between 1858 and ∼1926.

Individual authors

The entries below are ordered alphabetically by the last words of the authors' common pen/stage names, typeset in italics. The real/full name of the author (if different) and the years of her or his life are given in parentheses, followed by a brief nationality and profession descriptor. The remaining uncertainties are indicated with the tilde ($\sim$) and question signs.

---

[2]With deep apologies to other countries on both American continents, I use this common term for US-based authors.

[3]In view of the well-known "English-vs-British" dilemma, I use this term for all UK-based authors who lived after 1707, and cannot be clearly identified as either Northern Irish, or Scottish, or Welsh.

---

[4]He was also known as Cartesius – hence the Cartesian coordinates.

---

[5]He was King Edward VIII of the United Kingdom for a few months in 1936.

---

[6]Since 1983, he is also known as Dr. Fad.

---

[7]In some sources, referred to as John Andrew Holmes.

---

[8]Actually, this pen name had been used earlier by Ruth Crowley, who had started the *Ask Ann Landers* advice column in 1943; Lederer picked it up in 1955, and carried on until 2002.

[9]Sometimes (especially in French texts) spelled as Leibnitz.

---

[10]There is a (I believe, very small) chance that the actual author of that quote is another Dick Nolan, a Canadian musician Richard Francis Nolan (1939–2005).

[11]Sometimes spelled as Occam.

---

[12]He also wrote under pen names Bill Vaughan and Burton Hillis.

Printed in the United States
By Bookmasters